Investigating Animal Abundance:

capture-recapture for biologists

Michael Begon B.Sc. Ph.D.

Lecturer in Zoology, University of Liverpool

University Park Press

Baltimore

First published 1979
by Edward Arnold (Publishers) Limited, London

First published in the USA in 1979 by
University Park Press,
233 East Redwood Street,
Baltimore, Maryland 21202

Library of Congress Cataloging in Publication Data

Begon, Michael
Investigating animal abundance.

Bibliography: p.
1. Animal populations—Mathematical models
2. Animal marking I. Title
QH352.B44 1978 591.5′25 78–22544

ISBN 0–8391–1387–0

Printed in Great Britain

Contents

Preface

Ecology is a subject which, by its very nature, must face up to the complexities of the living world. It is, therefore, fraught with problems concerning the quantification of its subject matter. One of these problems is the estimation, or more generally the investigation, of animal abundance.

Finding the solution to this problem, as is so often the case, has largely been the work of mathematicians or statisticians. But the potential users of the solution are biologists. My aims, therefore, are to communicate to biologists, at both student and research level, the essential simplicity of the mathematical techniques; and also to reassure them that it is their experience, *as biologists*, which can turn a sterile mathematical technique into a potent biological weapon. Both of these aims are, of course, pertinent to the conjugation of mathematics and biology generally.

It is a pleasure to thank Dr. J. A. Bishop and Dr. K. O'Hara for their help with the manuscript; and my wife, Sally, for her help in clarifying passages which, despite my good intentions, I had left obscure. I should like, also, to acknowledge my debt to the paper by R. M. Cormack (1973), which first awakened me to the proposition that mathematical techniques can be reduced to common sense. I have tried, in what follows, to convey this same message.

Liverpool, 1978 M.B.

1 Context

Biologists all too often act as if mathematical techniques possess magical powers, transforming incomprehensible raw data into clear and precise scientific conclusions. Nothing could be further from the truth. Far from being magical, such techniques are, in fact, essentially sterile. They become potent only in the light of the biological problem to which they are applied, and the biological context in which they are interpreted. A consideration of mathematical techniques must, therefore, be preceded by a consideration of problems and contexts.

The roe-deer enjoys extreme popularity throughout most of Europe as an attractive member of the mixed-woodland community. Ironically, it is also a much-coveted hunting trophy. In 1962, for instance, Andersen reported that in Denmark alone around 25 000 roe-deer were being killed each year by more than 100 000 licensed sportsmen. Thus, the management of deer populations, and the policy regarding their shooting, must obviously be founded on a firm basis of biological fact. And especially important is the need for reliable information on the actual numbers of deer in particular woodlands (population size), or – what is essentially the same thing – knowledge of their absolute density (numbers per unit area).

Andersen described an attempt to measure the absolute density of roe-deer on the Danish Game Research Farm. The attempt was made by the forestry and game personnel, who had known the farm's woodlands for years, and spent part of every day there. No group could have been more experienced. They tried simply to count all the deer, and considered there to be about 70 present. But in the subsequent three months they managed to kill 213 deer in the same isolated woodland.

The experienced gamekeepers were incredulous at this discrepancy – but they should not have been. A simple count or census is justifiable – though not without difficulties – for an entirely visible and sedentary population. But roe-deer are continually, and unpredictably, on the move; and their ability to escape detection is obviously good enough to deceive even the most experienced census-taker. If a firm basis of biological fact is required for this mobile and cryptic mammal, then some more objective method of measuring absolute density is necessary.

Such information is urgently required – but only as a consequence of man's desire to exploit a part of his environment for his own pleasure. Frequently, however, the motivation behind population studies is much more humanitarian. Although nowadays we in the West are largely free from dangerous infectious disease, this is by no means world-wide. Dengue haemorrhagic fever, for instance, is a virus infection transmitted by the mosquito *Aedes aegypti*, which primarily affects young children. It was only recognized as a specific condition in 1953, but by the mid-1960's it had become a health problem in much of southern Asia.

In 1966, the World Health Organization established their Aedes Research Unit in Bangkok, the capital of Thailand. It had been noted there that the incidence of the disease showed a marked increase in the wet season, and that the outbreaks tended to be larger every second year. The presumption made by many researchers was that these cycles in the incidence of the disease were correlated with cycles in the density of the mosquitoes, or, perhaps, in their efficiency as vectors. The Aedes Research Unit set out to investigate whether there was any truth in these presumptions (Sheppard *et al.*, 1969). The aim was to control the disease. This could only be achieved by basing any plans on a proper understanding of mosquito ecology.

Among the relevant ecological parameters were the actual density of mosquito populations at different times of the year; the tendency of mosquitoes to move from one area (population) to another (which would affect the efficiency with which they located humans); and the life-expectancy of mosquitoes (which would also affect their efficiency as vectors). Life-expectancy can, of course, be quantified by measuring survival-rates: the higher a mosquito's chance of survival, the longer it can be expected to live.

It was essential that these parameters should be investigated under natural conditions: it was the dynamics of actual populations which were assumed to underlie the fluctuations in the disease's occurrence. The major problem facing the unit was, therefore, a practical one: to investigate the size, along with other parameters, of a population of actively mobile mosquitoes, in an urban environment large enough to support two million people – and to do this over an extended period of time, while the mosquitoes themselves passed through several generations: being born, maturing and dying.

Yet, if the ecology of the mosquitoes was to be properly understood, and an enlightened control plan instituted, this problem had to be solved. Progress towards the amelioration of dengue haemorrhagic fever was, therefore, crucially dependent on reliable information concerning the population parameters of the mosquitoes.

The mosquito and the roe-deer are both important to man. One we wish to control in order to alleviate suffering; the other we wish to exploit in order to increase pleasure. In both cases there is a requirement for information on the absolute density of non-cooperative populations, and on the processes leading to these densities: birth, death, emigration and immigration. Of course, such requirements are not confined to studies with an immediate application, as the following example shows.

Population genetics is essentially the study of the origin and dynamics of genetic variation within and between populations. In other words, the population geneticist seeks to account for the similarities and differences in gene frequencies between conspecific individuals, between conspecific populations, between closely-related species, and so on. In the past, there have been three classes of approach to this extremely daunting task: the theoretical, the laboratory and the field.

The theoretician identifies the processes which are potentially capable of affecting gene frequencies, and investigates their relative importance and combined action in a number of hypothetical circumstances. Many of these are themselves influenced by the ecological circumstances of the population in question. In spite of this, the theoretical population geneticist seeks to increase our understanding of the real world by considering idealized populations. This is the only way he can make any progress.

The population genetics fieldworker attempts to increase our understanding of the real world by studying the real world itself. Between him and the theoretician lies the laboratory worker. Laboratory work can uncover the potentialities of actual organisms. It can 'test' the models of the theoretician, and indicate to him what is biologically realistic; and it can provide the fieldworker with the results of controlled experiments to compare with any inference he may make. Ultimately, however, all population geneticists are interested in the real world. Yet, it is only by interpreting his results against a coherent theoretical background, that the fieldworker can hope to succeed in making the real world understandable. Fusing the different aspects of population genetics is obviously essential.

Towards this end the American geneticist Sewall Wright (1969) developed the concept of effective population size: essentially, the size of the ideal population with which an actual population can be equated genetically. In other words, by collecting the appropriate data from a natural population the effective population size can be estimated, and the results of the theoretician and laboratory worker applied to the field. As Wright himself has remarked, estimating effective population size is '. . . a practical necessity in dealing with natural populations'.

The quantification of absolute densities plays a crucial role in determining the effective size of a natural population. Yet the animals most often studied by population geneticists – butterflies, moths, fruit-flies, snails, and so on – are, once again, impossible to census by a direct count. It follows that the successful fusion of practical and theoretical aspects of population genetics is dependent on our ability to measure the absolute density of populations of animals, which are either mobile, cryptic or both.

It would be easy to follow example with example: the exploitation of freshwater fish for food; the destruction of verminous small mammals by poisoning; the protection of birds in disappearing parts of the environment; the removal of insect pests from crops; as well as the understanding of population dynamics for its own sake. All of these are areas where it is crucially important to measure the absolute density of populations of non-cooperative animals. Moreover, if we wish to exploit a fish population we will need to know the birth-rate; if we wish to destroy small mammals we will need to know the extent to which different populations intermix; and generally, if we wish to understand population dynamics we will need to study the forces underlying those dynamics: birth, death, immigration and emigration. Such an argument is easy to summarize: information on the absolute density of animal populations, and the forces determining density, is essential.

But the expressions–'information on' and 'measurement of' density – are unacceptably vague. The situation can be compared with a consideration of the length of the River Dee, or the volume of Lake Bala. Length, volume and number are all commonplace concepts, but there are cases (the River Dee, Lake Bala, and most animal populations) in which their measurement is by no means easy. Neither is it easy to say exactly where the River Dee begins or ends; or exactly what water level to assume in Lake Bala; or exactly where the limits of a population are. And finally, and by no means facetiously, for someone wishing to swim the Dee, it is only necessary to know if it is more or less than a mile or so long.

In other words, we must accept first of all that no population size can ever be specified exactly. But we must also accept that an exact specification is usually unnecessary. A geneticist measuring effective population size may only need to know whether his population is sufficiently small for genetic drift to be a potentially potent force. The Aedes Research Unit in Bangkok really only needed to know whether mosquito-density and disease-incidence were correlated. A Danish game-keeper may only need to know whether he should let 30 or 130 roe-deer be shot in a season. There are, of course, many situations where more precise information is required. But the basic point is the same: population sizes are measured in response to specific ecological questions, and the form of the question determines the precision and accuracy required of the answer.

In fact, one can go further. In an ideal world, precision and accuracy would always be maximized. But, in particular cases, time, money, personnel, and particularly technique may all be severely limited. In practice, therefore, precision and accuracy must be sacrificed in order to minimize the limitations of these other factors. The extent of the sacrifice is determined by the question in hand, and the circumstances of the study: context is all-important.

2 The Models

The first chapter established the interest of ecologists in estimating both the size of mobile animal populations and the strengths of those processes which determine the size. Operations designed to satisfy this interest will now be examined. All of the methods to be described involve marking, releasing and recapturing individuals. Most of them demand that these processes be repeated several times. A variety of generic names have been used for these methods: mark-release-recapture, release-recapture, multiple-recapture, and so on. I shall refer, in general, to capture-recapture methods and capture-recapture models.

The first chapter also established the importance of context in determining what methods should be used. This applies not only within the range of capture-recapture models, but also between capture-recapture and other models. There are undoubtedly situations in which the accuracy required of the answer, and the limitations in resources, suggest that some method other than capture-recapture is the most appropriate. Such methods will not be described here, because space is limited. No value judgement is implied.

2.1 The Petersen estimate

Even the most sophisticated models are directly descended from the simplest, which is the one first advocated by Petersen in 1896. It was also used by Lincoln in 1930 to estimate the size of the North American duck population, and is often called the Lincoln Index.

Imagine that we wish to estimate the size of a population into which there is neither birth nor immigration. On a first visit we catch a random sample of r individuals, mark them so that we can recognize them in future, and return them to the population. They remix perfectly with the unmarked individuals. Subsequently there is both death and emigration, to which, however, marked and unmarked animals are equally prone. The marked *proportion* remains the same as when the r marked individuals were initially released. On a second visit a further random sample is caught: total size n, of which m individuals are marked. If the size of the whole population immediately before the first visit was N, then it should be true that:

$$\frac{m}{n} = \frac{r}{N}$$

i.e. the marked proportion has remained the same, and our random sample of a perfectly mixed population reflects this.

We can now estimate N. The symbol $\hat{N}$ ('N-hat') denotes 'an estimate of N', and therefore:

$$\hat{N} = \frac{rn}{m}$$

This is the Petersen estimate.

Imagine that the population *is* subject to birth and immigration, but *not* to death and emigration. After the first visit, all of the r marked individuals would remain in the population, but there would be neither birth nor immigration of marks. The number of unmarked individuals, however, would increase steadily and the marked proportion steadily decline. The marked proportion in the second random sample would, therefore, reflect the situation *at that time*. In other words, $\hat{N}$ would refer not to the first, but to the second visit.

Of course, if there is neither birth, death, emigration nor immigration, the population size remains constant, and $\hat{N}$ refers to both first and second visits. In fact, this (most restrictive) case is the one usually envisaged for the Petersen estimate. For convenience it will be referred to as the 'simple Petersen estimate'.

In 1951, Bailey showed that, in cases where the numbers involved were small (m around 10, or less), the modified formula:

$$\hat{N} = \frac{r(n+1)}{(m+1)}$$

gave a more accurate (less biased) estimate. When numbers are large, the difference between the modified and unmodified formulae is negligible. The modified formula is, therefore, of more general application. Similar formulae will be used in most future calculations.

Bailey also derived a formula for the standard error of this estimate:

$$SE_{\hat{N}} = \sqrt{\frac{r^2(n+1)(n-m)}{(m+1)^2(m+2)}}$$

The Petersen estimate is the simplest estimator of population size using marked individuals, but it is also the one of most restricted utility. There are two basic reasons for this. The first is that it involves just one release and one recapture; even the simplest alternatives improve upon it by using more data. The second is that there are marginally more assumptions implicit in the Petersen estimate than in most other models. These assumptions must now be examined.

2.2 Capture-recapture assumptions

1) The first, and almost trivial, assumption is that all marks are permanent, and are noted correctly on recapture. This, of course, refers only to the period of study: subsequent losses are irrelevant.
2) The second assumption is that being caught, handled and marked one or more times has no effect on an individual's subsequent chance of capture. This infers both that the inherent 'catchability' of an individual is unaffected by being caught; and that the position of marked individuals in the population, after sampling, is no different to that which would be expected if they had never been caught.
3) Thirdly, it is assumed that being caught, handled and marked one or more times has no effect on an individual's chances of dying or emigrating. Implicit in this assumption is another one: that all emigration is permanent, and therefore essentially indistinguishable from death.
4) Furthermore, it is assumed that all individuals–whether marked or not–have, inherently, an equal chance of being caught. This is tantamount to assuming that the population is sampled at random, without regard for the age, sex or physiological condition of individuals. Note that this assumption will still hold if, on a particular day, catchabilities are not equal, but individuals are assigned a catchability-class at random. In other words we are assuming that there is no *inherent* difference, *not* that on a particular day there is no difference at all. It follows from this that individuals from different classes within the population will be sampled in the proportion in which they occur.

   The analysis of a heterogeneous population will produce results which are applicable neither to the individuals or individual classes, nor to the population as a whole. Consequently this assumption can, and should whenever possible, be side-stepped by dealing with the different sexes, age-classes, etc. separately.
5) It is also assumed that all individuals – whether marked or not – have, inherently, an equal chance of dying or emigrating.

These five assumptions apply to almost all capture-recapture models. The exceptions will be indicated in the models concerned.

6) The Petersen estimate assumes either that there are no births or

immigrations, or that there are no deaths or emigrations, or that there are none of these. Most alternative models do not make any of these assumptions.

7) The final assumption – which applies to all models that do *not* assume that there is neither birth, death, emigration nor immigration – is that sampling periods are short in relation to total time. This is because birth, death, immigration and emigration are all processes which we may call quasi-continuous. That is to say we treat them as continuous, while recognizing that they are actually made up of a series of discrete, singular events. We know, for instance, that death affects one whole individual at a time, but we might still compute a death-rate of 0.05 individuals per minute. In other words, we recognize that there is a continuous *possibility* of death.

   Most capture-recapture models include these quasi-continuous processes, and many actually quantify them. If they are to be quantified, then they must be measured between two points in time. This means that sampling periods are assumed to *be* points in time. In fact, even if the processes are not quantified, population size is estimated for a sampling occasion on the assumption that the processes do not alter population size *during* the sampling occasion. Once again, a sampling period is assumed to be a point in time. Strictly speaking, this assumption can never hold – sampling can never be instantaneous – but attempts to conform as closely as possible to the assumption should still be made. Sampling periods should be short in relation to total time. This should lead, in turn, to the intervals between samples being discrete.

2.3 Notation

Before the individual models are examined, it will be useful to describe a common notation which will be employed throughout.

2.3.1 Sampling

In essence, all of the following methods involve taking a series of samples from the population under study. Most commonly, the intervals between these samples are one or more days, and sampling is said to occur on 'day *1*', 'day *2*' etc. Remember that a 'day *1* sample' refers to a sample taken over a short period during day *1*. Remember, also, that other time intervals are equally acceptable, and that 'days' refer to any discrete time interval. It follows from this that in practice it is always advisable, and in some methods essential, for samples to be taken at the *same* time each day. In this way all time intervals are the same, or, at worst, simple multiples of one another.

The size of each sample, the number caught, will be denoted by n. Thus, on day *1* n_1 individuals are caught; on day *2* n_2 individuals are caught; and, in general, on day i n_i individuals are caught.

Often, every one of these individuals will be marked and released. Sometimes, however, animals are harmed or damaged during handling, so that the number of marked individuals released on day i is less than the number originally caught (n_i). The number of marked individuals released on day i will be denoted by r_i.

2.3.2 *Marking*

A brief comment about marking is convenient here, although this will be examined again in Chapter 5. There are, essentially, three types of mark. The first is individual-specific, allowing each animal to be recognized individually, and providing the maximum amount of information on recapture. Such information is, however, largely superfluous in the present context. The second type is date-specific. On recapture, such marks allow the previous occasion or occasions on which the animal was caught to be noted. A single, individual-specific mark does, of course, have this same capacity. The third type of mark is neither individual- nor date-specific. It merely allows animals to be classified as marked or unmarked, providing the minimum amount of information.

Most of the following methods presume that marking is date-specific, and such marking is, therefore, to be recommended. When marking is individual-specific, the pattern of marks on each recaptured animal that *would* have resulted from date-specific marking must be imagined. This imaginary pattern must then be used.

On every day except the first a proportion of the sample will probably be marked. In those cases where the total number of marks caught on day i is required, this will be denoted by m_i. Often, however, it will also be necessary to partition m_i according to when the mark was given. For instance, of m_4 marks caught on day *4*, some will be from day *1*, some from day *2*, and some from day *3*. These will be denoted by $m_{4\,1}$, $m_{4\,2}$ and $m_{4\,3}$ respectively; and, in general, by $m_{i\,j}$ where i is the day of capture, and j the day of marking.

2.3.3 *Population size*

The most obvious, but by no means the sole, purpose of capture-recapture studies is the estimation of population size itself. The population size on day i will be referred to as N_i.

2.3.4 Death and emigration

Capture-recapture methods do not, in themselves, distinguish between death and emigration. They must, therefore, be considered together as 'loss'. L_i will be used to denote the *number* of individuals lost from a population between days i and $i+1$. (For instance, L_2 is the number of individuals either dying or emigrating between days 2 and 3.)

Those individuals which are not lost survive, and it is often survival rather than loss which is measured. This, apart from anything else, avoids the disadvantage of equating death and emigration. In fact it is usual to consider survival-*rate*: the proportion of the population surviving from one occasion to the next, or the probability of any one individual surviving from one occasion to the next. ϕ_i ('phi'$_i$) will be used to denote the proportion surviving from day i until day $i+1$. Several methods calculate one survival-rate only, on the assumption that the rate of survival is constant. This will be denoted simply by ϕ.

2.3.5 Birth and immigration

Capture-recapture methods are also unable to distinguish between birth and immigration, and these too must be considered together – as 'gains' or 'additions'. B_i will denote the number of additions between days i and $i+1$, and b_i will denote the proportion of the day $i+1$ population that were added between days i and $i+1$. When b_i is assumed to remain constant from day to day it will be referred to as b.

2.3.6 'Marks at risk'

The simplest Petersen estimate rests on the assumption that the marks released on the first occasion represent the sum total of marks available for recapture on the second. All alternative methods, however, modify this assumption to some extent. Primarily, it is generally recognized that a proportion of the marks are subject to either death or emigration each day. But the consequent decline in the number of marks is usually opposed by the fact that fresh marks are regularly being added to the population.

These two factors combine to determine the 'marks at risk', M_i: the number of marks in the population which are available for sampling immediately before the day i sample. Obviously, additions to M_i can only be made with the experimenter's knowledge – by releasing a known number of marked individuals. Yet in other respects M_i is assumed to be a true sub-population of N_i: loss-rates in the marked and unmarked portions of the population are assumed to be the same, proportions captured of the marked and unmarked portions are assumed to be the same, and so on. In other words, the marked individuals are regarded as truly representative of the whole population. Measurements can be made on those marked individuals, because they are identifiable. These measurements are assumed to be equally applicable to the whole population.

It should be clear that M_i is analogous to the term r in the Petersen estimate. Not surprisingly, the estimation of M_i is a crucial part of most Petersen estimate derivatives.

2.3.7 *Summation*

Most of the terms introduced so far have had two components: the quantity which they measure, and the day to which they apply. For instance, m_3 refers to the number of marks caught – on day *3*. Many of the following methods require several of these terms to be added together. For instance, in a study lasting five days we may need to know the following sum: $r_1 + r_2 + r_3 + r_4$ – the total number of marked individuals released (none are released on day *5*). It is convenient to have a shorthand method of representing such sums, and the one generally used is:

$$\sum_{i=1}^{i=4} r_i$$

Put into words, this is the sum of the r_i's, where i takes all values from *1* to *4*.

Furthermore, in this particular case, since we know that marked individuals are released on all days except the last, it is really only necessary to write:

$$\sum_i r_i \text{ or even } \sum r_i$$

– the total number of marked individuals released.

The notation used in the models is tabulated in Table 2.1.

We are now in a position to consider the various capture-recapture models. In each case, the model itself will be described first, followed by a worked example, and then by a discussion of the model's utility. The description will concentrate on rationale. This should promote a critical appreciation of the model, but may do so at the expense of a superficial and rapidly-learnt ability to use it. For this reason, it is likely that an understanding of how to apply the model – as opposed to an understanding of the model itself – will be developed by considering the description and worked example together.

Many of the models include formulae for calculating the standard error of their estimates. Standard errors measure precision, and, at

Table 2.1 Summary of notation.

a_i	the number of individuals caught both on day i and day $i+1$
b_i	the proportion of the day $i+1$ population added between days i and $i+1$
m_i	the number of marked individuals caught on day i
m_{ij}	the number of individuals caught on day i with a day j mark
n_i	the number of individuals caught on day i
p_i	the proportion of the population captured on day i
q_i	the proportion of the day i sample carrying a mark
r_i	the number of (marked) individuals released on day i
y_i	the number of individuals marked (and released) on day i and caught again subsequently
z_i	the number of individuals marked before day i, not caught on day i, but caught again subsequently
A_i	the mean age of marks on day i
B_i	the number of additions to the population between days i and $i+1$
L_i	the number of losses from the population between days i and $i+1$
M_i	the number of marks at risk on day i
M_{ij}	the number of day j marks at risk on day i
N_i	the population size on day i
T_i	the total age of all marks on day i
W_i, X_i, Y_i, Z_i	mutually exclusive subsets of the day i population (see Manly and Parr's method).
$\alpha_i, \beta_i, \gamma_i, \delta_i$	the number of immigrations, births, emigrations and deaths respectively in the population between days i and $i+1$
ϕ_i	the proportion of the day i population surviving until day $i+1$; or the chances of an individual in the day i population surviving until day $i+1$

face value, the inclusion of these formulae seems to be an unquestionable advantage. There are, however, good reasons for believing that the increase in statistical reliability which these standard errors imply is, in fact, spurious; this will be examined more fully in the next chapter. In order to include these formulae, but to make their inclusion unobtrusive, I have placed them–without discussion–in the *Examples*.

Computations in some of the models are lengthy, and the routes to the final estimates are occasionally tortuous. At no time in the discussions has this been considered a disadvantage. Even the longest calculations will be considerably shorter than the data-collecting leading up to them. If a model is appropriate then time should be of no consequence.

2.4 Weighted mean

This method differs very little from the simple Petersen estimate, and makes all the same assumptions. The major difference is that it utilizes data collected over several days. A quite different derivation of the model has been described by Seber (1973, p. 139). On each day, n_i

individuals are caught, of which m_i are already marked. Unmarked individuals are then marked, and r_i marked individuals released. There are, therefore, $(r_i - m_i)$ additional marked individuals released each day.

Since we are assuming that there are no losses from the population, the number of marked individual increases with time. The number of marks at risk on day i – M_i – is, therefore, the number of marks released on day *1* (r_1) plus the number of additional marks released on all days between day *1* and day i i.e. $((r_2 - m_2) + (r_3 - m_3) \ldots + (r_{i-1} - m_{i-1}))$. Population size can then be estimated using a standard Petersen estimate:

$$\hat{N}_i = \frac{M_i \, n_i}{m_i}$$

We have assumed, however, that the population is closed, and that all $\hat{N}_i$'s are themselves estimates of the same thing: N. Yet some $\hat{N}_i$'s are likely to be more accurate than others, and it is undoubtedly sensible to give most weight to the most accurate estimates. To do this we must calculate a weighted mean of $\hat{N}_i$'s, which – if we denote the weight attributed to $\hat{N}_i$ as w_i – is given by:

$$\hat{N} = \frac{\Sigma N_i w_i}{\Sigma w_i}$$

We now need some actual values for the w_i's. These can be derived simply by arguing that the accuracy of an $\hat{N}_i$ depends on how much information it was derived from. In other words, an $\hat{N}_i$ obtained from a sample containing very few marked individuals is likely to be much influenced by chance effects on the actual size of m_i. Samples in which m_i is large, on the other hand, will be much less influenced by chance effects, and will, on average, produce more accurate $\hat{N}_i$'s. We can, therefore, use the m_i's as our weights:

$$\hat{N} = \frac{\Sigma \hat{N}_i m_i}{\Sigma m_i}$$

This, because of the formula used to calculate $\hat{N}_i$, reduces to:

$$\hat{N} = \frac{\Sigma M_i n_i}{\Sigma m_i}$$

A virtually unbiased form of which is:

$$\hat{N} = \frac{\Sigma M_i n_i}{(\Sigma m_i) + 1}$$

Example. The following data were collected from a grid of small-mammal traps for a population of the wood mouse *Apodemus sylvaticus*:

day (i)	*1*	*2*	*3*	*4*
n_i	–	6	4	7
m_i	–	1	2	2
r_i	8	6	4	–

This table can be continued:

M_i	–	8	13	15
$M_i n_i$		48	52	105

$$\text{and } \hat{N} = \frac{(48+52+105)}{(1+2+2)+1} = 34.2$$

The standard error is given by

$$\mathrm{SE}_{\hat{N}} = \hat{N}\sqrt{\frac{1}{\Sigma m_i+1}+\frac{2}{(\Sigma m_i+1)^2}+\frac{6}{(\Sigma m_i+1)^3}} = 17.1$$

The estimated size of our wood mouse population is 34.

The weighted mean model assumes as much as the most restrictive Petersen estimate: that the population is closed, and has neither births nor deaths. (In fact, for the population in the *example*, these assumptions are unlikely to be strictly true). Its only advantage over the Petersen estimate is that it accumulates data over several days. Yet this advantage may be considerable. In studies where there are very few captures, and application of the more sophisticated models is impossible, we may be left with the choice of using several Petersen estimates, using a weighted mean, or abandoning the whole venture. The last alternative is really only appropriate when the restrictive assumptions are very far from the truth – otherwise it is unnecessarily wasteful. If several Petersen estimates are used, there is a further choice to be made. Either the variations in size are accepted as real – which, in view of the scanty data, is inadvisable; or the various estimates are used to produce a single mean or average. This is unlikely to compare favourably with our weighted mean derived from a comparatively large number of marks.

Of course, there may be good reasons for believing that there is loss but no gain, or gain but no loss. The appropriate Petersen estimate may then be used. It is unlikely, however, that there will be emigration but no immigration, or *vice versa*. Use of the Petersen estimate with several days' data is, therefore, restricted to cases where there is birth only or death only. Thus the Petersen estimate makes only marginally fewer assumptions than the present method which – with the advantages of a weighted mean and accumulated data – will often be the best way of extracting an estimate from otherwise inadequate data.

Overall, it is inadvisable to *plan* to estimate population size by the weighted mean method unless the answer is of absolutely crucial importance, and no possible alternative exists. But if the fieldwork has been done and the data are poor, this method, interpreted cautiously, is often better than nothing at all.

2.5 Jackson's positive method

The principle of this method is that marking occurs on one occasion only, followed by several recaptures. It is very much a derivative of the 'positive' method suggested by Jackson in 1937. It is also very similar to a method described by Seber (1973, p. 260). On day *0*, r_0 individuals are caught, marked and released, and on subsequent days n_i individuals are caught, of which m_i are marked. These individuals may be released again, but no further marking takes place after day *0*.

q_i is the proportion of the day i sample that are marked:

$$q_i = \frac{m_i}{n_i}$$

Although there are losses from the population after the day *0* sample, this should not affect the *proportion* of marks in either the whole population or in future samples – marked and unmarked individuals should die and emigrate at the same rate. There are additions, however, to the unmarked portion of the population, but none to the marked portion. q_i should, therefore, decline with time.

Our aim is to estimate N_0, and this can be achieved by estimating q_0: the marked proportion of a hypothetical random sample taken on day *0*, before any additions have been made to the population. The marked proportion in this hypothetical sample should be the same as in the population:

$$q_0 = \frac{r_0}{N_0}$$

and, therefore:

$$\hat{N}_0 = \frac{r_0}{q_0}$$

q_0 can be estimated as follows. Define b as the birth-rate per day in the population. On, for instance, day *3* we can divide the population into two classes: marked and unmarked. q_3 is an estimate of the marked proportion. We can divide the unmarked class into two further sub-classes: absent on day *2*, and present on day *2*. By definition, b is the proportion of the total that are in the first of these sub-classes: present and unmarked on day *3* but absent on day *2*.

The remaining sub-class–$(1-b-q_3)$ of the total population–contains unmarked individuals that were also in the population on day *2*. These, like the marked individuals, have survived from the day *2* population where their relative proportions were $(1-q_2)$ and q_2. Moreover, their survival rates from day *2* to day *3* should be equal, and these relative proportions should, therefore, stay the same:

$$\frac{q_2}{1-q_2} = \frac{q_3}{1-b-q_3}$$

This rearranges to:

$$q_3 = q_2(1-b)$$

Similarly: $$q_4 = q_3(1-b) = q_2(1-b)^2$$

and generally: $$q_i = q_0(1-b)^i$$

or $$\ln q_i = i(\ln(1-b) + \ln q_0)$$

This is the regression equation of $\ln q_i$ on i (both of which are known), and the two regression constants—$\ln(1-b)$ and $\ln q_0$–can be calculated. Thus $\hat{q}_0$ (and subsequently $\hat{N}_0$) and b can themselves be computed. It would be sensible, however, to recognize that the q_i-values (and thus the $\ln q_i$-values) are not equally accurate. Those derived from smaller m_i's are more susceptible to chance effects. m_i-values can, therefore, be used as weights in the calculations:

$$\ln(1-b) = \frac{\Sigma m_i(\ln q_i - \overline{\ln q})(i-\bar{i})}{\Sigma m_i(i-\bar{i})^2}$$

$$\ln q_o = \overline{\ln q} - \ln(1-b)\bar{i}.$$

Example. The following data are those reported by Jackson (1939) himself for a population of tsetse flies which he sampled at weekly intervals:

week (i)	*0*	*1*	*2*	*3*	*4*	*5*	*6*
r_i	1558						
n_i		1547	1307	603	1081	1261	1798
m_i		101	81	24	31	10	11

This table can be continued:

$q_i = m_i/n_i$	0.0653	0.0620	0.0398	0.0287	0.0079	0.0061
$\ln q_i$	−2.729	−2.781	−3.224	−3.552	−4.873	−5.097

Giving:

$$\ln(1-b) = \frac{-419.218}{918.50} = -0.456$$

$$\text{or } \hat{b} = 0.366$$

$$\text{and } \ln q_0 = -3.709 - (-0.456)\, 3.5 = -2.113$$

$$\text{or } \hat{q}_0 = 0.121$$

$$\text{and } \hat{N}_0 = \frac{1558}{0.121} = 12889$$

The standard errors can be obtained from the weighted regression:

$$\text{SE}_{\ln(1-b)} = \sqrt{\frac{\Sigma m_i\{\ln q_i - \overline{\ln q} - \ln(1-b)(i-\bar{i})^2\}}{(n-2)\ \Sigma m_i(i-\bar{i})^2}} = 0.159$$

$$\text{SE}_{\ln q_0} = \sqrt{\frac{\Sigma m_i(\ln q_i - \overline{\ln q} + \bar{i}\ln(1-b))^2}{(n-1)\Sigma m_i}} = 0.515$$

Both of these standard errors need to be anti-logged, and this leads to the final standard errors being asymmetrical:

the values one standard error from $\hat{b}$ are 0.257 and 0.459;
the values one standard error from $\hat{N}_0$ are 21 580 and 7701.

Jackson's positive method requires several visits to the study site, but provides only one estimate of population size, and one of gain-rate – which is additionally assumed to remain constant throughout. The main improvement on simpler estimators is that loss, at a variable rate, is allowed. There is, however, relatively little return, in terms of estimates, for the effort expended. This is a result of the fact that individuals are marked only once – which is the major drawback of the method, but also its major advantage.

The drawback aspect is obvious: a single marking occasion leads to only one estimate, which represents a relatively small amount of information per unit effort. The advantage of the method is that it is ready-made for situations in which animals *can* only be marked on a single occasion. For instance, insects may sometimes be marked by giving them radioactively-labelled food (a single marking occasion), after which subsequent samples must be screened in a laboratory. Also, fish may be marked on a single occasion, and trawlers relied on for (weekly) recaptures. Moreover, it is possible to imagine a situation in which marking is a difficult procedure, and substantial trained assistance is available only once. A single mass-marking occasion can, then, be followed by several recaptures using untrained assistance. In such circumstances, Jackson's positive method makes efficient use of the available data and may be extremely useful. It is also possible to use this method when there are very few marking occasions, rather than only one. Generally, however, these situations are probably rather rare.

Note that, because this method depends on proportions, samples can be removed from the population (for screening) and never replaced. Note also the connection between this and other models: it is an extension of the Petersen estimate which allows loss but no gain and estimates population size for the first sampling occasion. In the present case, loss has again been 'allowed', but gain has been presumed constant and estimated.

2.6 The triple catch method

This method requires only three sampling occasions, but allows a comparatively large number of parameters to be estimated. Exactly how many depends on what assumptions are made.

On day *1*, a sample is taken and r_1 marked individuals are released. On day *2*, n_2 individuals are caught of which $m_{2\,1}$ are marked. All r_2 undamaged individuals, whether previously marked or unmarked, are then marked and released. On day *3*, n_3 individuals are caught of which $m_{3\,1}$ bear a day *1* mark only, and $m_{3\,2}$ a day *2* mark; individuals bearing both marks are included only in the $m_{3\,2}$.

Initially, we shall allow birth- and survival-rates to vary from day to day. On day *2*, there are $M_{2\,1}$ marks at risk. If the marked proportion is the same in the sample and the population, then:

$$\frac{M_{2\,1}}{N_2} = \frac{m_{2\,1}}{n_2}$$

and

$$\hat{N}_2 = \frac{M_{2\,1} n_2}{m_{2\,1}}$$

Also, since these $M_{2\,1}$ individuals are the survivors of the r_1 individuals released the day before:

$$M_{2\,1} = r_1 \phi_1$$

and
$$\hat{\phi}_1 = \frac{M_{2\,1}}{r_1}$$

What we require is an estimate of $M_{2\,1}$.

During the day *2* sample, $m_{2\,1}$ marked individuals are removed. Immediately after this sample there are, therefore, $(M_{2\,1} - m_{2\,1})$ day *1* marks still at risk. There are also r_2 day *2* marks at risk at this time, because they have just been released. The survival-rates of these two groups until day *3*, and the proportions of them taken in the day *3* sample ought to be the same. That is:

$$\frac{m_{3\,1}}{M_{2\,1} - m_{2\,1}} = \frac{m_{3\,2}}{r_2}$$

and therefore:

$$\hat{M}_{2\,1} = \frac{m_{3\,1} r_2}{m_{3\,2}} + m_{2\,1}$$

Finally, b_2 – the proportion of the day *3* population which had been added between days *2* and *3* – can be estimated by a procedure analogous to that used in Jackson's positive method. There it was shown that:

$$q_3 = q_2(1 - b_2)$$

In the present case q_2 and q_3 are represented by $m_{2\,1}/n_2$ and $m_{3\,1}/n_3$, and so

$$\hat{b}_2 = 1 - \frac{m_{3\,1} n_2}{n_3 n_{2\,1}}$$

All that remains is for allowance to be made for bias, leading to the following equations:

$$\hat{M}_{2\,1} = \frac{m_{3\,1}(r_2 + 1)}{(m_{3\,2} + 1)} + m_{2\,1}$$

$$\hat{N}_2 = \frac{(n_2 + 1)\hat{M}_{2\,1}}{(m_{2\,1} + 1)}$$

$$\hat{\phi}_1 = \frac{\hat{M}_{2\,1}}{r_1}$$

and
$$\hat{b}_2 = 1 - \left[\frac{(m_{3\,1} + 1)n_2}{(n_3 + 1)m_{2\,1}}\right]$$

This model, as it stands, is a special (three-day) case of Jolly's method (see below). We can only obtain further estimates by making the additional assumptions that both birth-rate and survival-rate remain constant. If the interval between days *1* and *2* (t_1) is the same as that between days *2* and *3* (t_2), this means that:

$$\hat{\phi}_2 = \hat{\phi}_1$$

and
$$\hat{b}_1 = \hat{b}_2$$

It may be, however, that t_1 and t_2 are not the same, in which case:

$$\hat{\phi}_2 = \hat{\phi}_1^{\,t_2/t_1}$$

and
$$\hat{b}_1 = 1 - \left[\frac{(m_{3\,1}+1)n_2}{(n_3+1)n_{2\,1}}\right]^{t_2/t_1}$$

We now argue as follows: of the day *2* population, $N_2\phi_2$ survive until day *3*. Since b_2 is the proportion of N_3 which were added between days *2* and *3*, $1-b_2$ must be the proportion which were not added. In other words $N_2\phi_2$ represents $1-b_2$ of N_3:

$$N_2\phi_2 = (1-b_2)N_3$$

Thus,
$$\hat{N}_3 = \frac{\hat{N}_2\hat{\phi}_2}{1-\hat{b}_2}$$

And by analogous reasoning:

$$N_1\phi_1 = (1-b_1)N_2$$

and
$$\hat{N}_1 = \frac{(1-\hat{b}_1)\hat{N}_2}{\hat{\phi}_1}$$

We have, therefore, derived estimates for N_2, ϕ_1 and b_2 which allow birth and survival-rates to vary; and additional estimates for N_1, N_3, ϕ_2 and b_1 on the assumption that these rates are constant.

Example. The following data refer to females of the grasshopper *Gomphocerippus rufus:*

day:	*1*	*2*	*3*
captured:		$50(n_2)$	$38(n_3)$
day *1*-marked:		$11(m_{2\,1})$	$5(m_{3\,1})$
day *2*-marked:			$9(m_{3\,2})$
released:	$42(r_1)$	$50(r_2)$	

One of the day *2*-marked individuals also had a day *1* mark, but this has been ignored.

Substituting into the equations:

$$\hat{M}_{21} = \frac{5 \times 51}{10} + 11 = 36.5$$

$$\hat{N}_2 = \frac{51 \times 36.5}{12} = 155.1$$

$$\hat{\phi}_1 = \frac{36.5}{42} = 0.87$$

$$\hat{b}_2 = 1 - \frac{6 \times 50}{39 \times 11} = 0.30$$

By assuming that birth- and survival-rates remained constant, and in the knowledge that the intervals between samples were the same:

$$\hat{\phi}_2 = \hat{\phi}_1 = 0.87$$

$$\hat{b}_1 = \hat{b}_2 = 0.30$$

$$\hat{N}_3 = \frac{155.1 \times 0.87}{0.7} = 192.8$$

$$\hat{N}_1 = \frac{0.7 \times 155.1}{0.87} = 124.8$$

Finally, the following standard errors can be calculated:

$$\mathrm{SE}_{\hat{N}_2} = \sqrt{\hat{N}_2(\hat{N}_2 - n_2)\left\{\frac{\hat{M}_{21} - m_{21} + r_2}{\hat{M}_{21}}\left(\frac{1}{m_{32}} - \frac{1}{r_2}\right) + \frac{1}{m_{21}} - \frac{1}{n_2}\right\}} = 64{\cdot}7$$

$$\mathrm{SE}_{\phi_1} = \sqrt{\hat{\phi}_1^2\left\{\frac{(\hat{M}_{21} - m_{21})(\hat{M}_{21} - m_{21} + r_2)}{\hat{M}_{21}^2}\left(\frac{1}{m_{32}} - \frac{1}{r_2}\right) + \frac{1}{\hat{M}_{21}} - \frac{1}{r_1}\right\}} = 0.$$

Making comparatively few assumptions we could say that:
The estimated population size for G. rufus *females on day* 2 *was 155; their estimated survival-rate between days* 1 *and* 2 *was 0.87; and their estimated gain-rate between days* 2 *and* 3 *was 0.30.*

If we assume that these rates remained constant, we can also say that:
The population size of G. rufus *females rose steadily from 125 on day* 1 *to 193 on day* 3.

The triple catch method, as presented here, is really two methods in one. The first provides just one estimate of population size, one of survival-rate, and one of gain-rate. But it does so after only three visits to the population, and it allows both gain- and loss-rates to vary. In other words, the method makes comparatively few assumptions, while producing a considerable amount of information per unit effort.

The second method provides even more information, but does so at the expense of making extra assumptions. Note, for instance, that the trend in population size uncovered in the example, is a result, purely and simply, of our assumption that gain- and loss-rates are constant. Such 'trends' will always accompany this second method. It is probably wisest, therefore, to be satisfied with the three estimates produced by the first method, unless there is a special reason for wanting the extra information of the second.

Time is often limited. It may be desirable, for instance, to compare several populations during a short period. Whenever this is so, the triple catch method will be an extremely quick, simple and efficient method of estimation.

2.7 Jackson's negative method

In complete opposition to his positive method, Jackson's negative method requires marked samples of known size to be released on several days, but the marks in the sample to be determined on the final day only. The model described here is a modification of Jackson's (1937) original method proposed by Bailey (1951).

Imagine a total of five sampling occasions. On days *1* to *4* random samples are marked and released (r_1 to r_4). On day *5*, n_5 individuals are caught of which $m_{5\,1}$ bear a day *1* mark, $m_{5\,2}$ a day *2* mark, and so on. A total of $m_5 (= \sum_j m_{5\,j})$ marks are caught on day *5*.

Assume a constant survival-rate, ϕ. Of the r_4 marks released on day *4*. $r_4\phi$ survive to be at risk on day *5*. Similarly, $r_3\phi^2$ day *3* marks are at risk on day *5*, and so on. A total of $M_5 (= \sum_j r_j \phi^{(5-j)})$ marks are at risk on day *5*.

The ratio of marks to individuals in the day *5* sample should be the same as in the whole day *5* population:

$$\frac{m_5}{n_5} = \frac{\sum_j r_j \phi^{(5-j)}}{N_5}$$

Our aim is to estimate N_5, but, unfortunately, in the above equation both ϕ and N_5 are unknown. In order to estimate N_5 we must, therefore, estimate ϕ first. This can be done by comparing the observed mean age of marks in the day *5* sample, with an estimated mean age of marks in the whole day *5* population.

Consider the sample first. Each mark is $(5-j)$ days old. The mean age of marks in the sample is, therefore, given by:

$$\frac{\text{total age of all marks}}{\text{total number of marks}} = \frac{\sum_j m_{5j}(5-j)}{m_5}$$

In the population, the total age of all marks on day *5*, T_5, is similarly given by:

$$\hat{T}_5 = \sum_j r_j \phi^{(5-j)}.(5-j)$$

and the expected mean age of marks in the whole population is $\hat{T}_5/\hat{M}_5$. This should, of course, be equal to the mean age actually observed in the day *5* sample:

$$\frac{\sum_j m_{5j}(5-j)}{m_5} = \frac{\sum_j r_j \phi^{(5-j)}.(5-j)}{\sum_j r_j \phi^{(5-j)}}$$

ϕ is the only unknown in this equation, but it cannot be obtained directly. The left-hand side of the equation is known, however, and every value of ϕ will lead to a different value for the right-hand side. The value of ϕ that leads to the left- and right-hand sides being equal is the one we require. Therefore, we plot a graph of ϕ against 'expected mean age', and read from it the ϕ-value associated with our observed mean age. Four points, suitably positioned, are sufficient.

Finally, once this ϕ-value has been obtained, N_5 can be estimated by rearranging the initial equation:

$$\hat{N}_5 = \frac{n_5}{m_5}.\hat{M}_5$$

or, following Bailey (1951):

$$\hat{N}_5 = \frac{\hat{M}_5(n_5+1)}{(m_5+1)}$$

Example. Once again, the tsetse fly data reported by Jackson (1939) can be used:

week (j)	*1*	2	3	4	5	6	7
r_j	1262	1086	1299	1401	1198	1183	
n_j							1558
m_{7j}	1	5	17	28	48	70	

This table can be continued:

week (j)	*1*	*2*	*3*	*4*	*5*	*6*	*7*
$m_{7j}(7-j)$	6	25	68	84	96	70	

Thus:

$$\begin{array}{l}\text{expected}\\ \text{mean age}\end{array} = \frac{\sum_j m_{7j}(7-j)}{m_7} = \frac{6+25+\ldots+70}{1+5+\ldots+70} = \frac{349}{169} = 2.065$$

We need to know the ϕ value that will give an expected mean age of 2.065. Using a trial ϕ of 0.5:

week (j)	*1*	2	3	4	5	6
$\phi^{(7-j)}$	0.0156	0.0313	0.0625	0.125	0.25	0.5
$r_j\phi^{(7-j)}$	19.719	33.999	81.188	175.125	299.5	591.5
$r_j\phi^{(7-j)}.(7-j)$	118.314	169.995	324.75	525.375	599.0	591.5

and,

$$\begin{array}{l}\text{observed}\\ \text{mean age}\end{array} = \frac{\hat{T}_7}{\hat{M}_7} = \frac{\sum_j r_j\phi^{(7-j)}(7-j)}{\sum_j r_j\phi^{(7-j)}} = \frac{2328.934}{1201.031} = 1.939$$

The 'true' ϕ is, therefore, greater than 0.5.

A ϕ of 0.6 leads, by similar calculations to an expected mean age of 2.238. The true ϕ, therefore, lies between 0.5 and 0.6, and further calculations give:

ϕ	expected mean age
0.53	2.026
0.56	2.115

The (virtually straight-line) graph of ϕ against expected mean age can now be drawn (Fig. 2.1). This indicates that an expected mean age of 2.065 would be obtained from a ϕ-value of 0.543.

It is sensible to check that this is correct – especially as some of the computations involved in this check are also necessary when estimating N_7:

$$\phi = 0.543$$

week (j)	*1*	2	3	4	5	6
$\phi^{(7-j)}$	0.0256	0.0472	0.0869	0.1601	0.2948	0.543
$r_j\phi^{(7-j)}$	32.349	51.259	112.883	224.300	353.170	642.369
$r_j\phi^{(7-j)}.(7-j)$	194.093	256.296	451.532	672.900	706.341	642.369

$$\text{expected mean age} = \frac{\hat{T}_7}{\hat{M}_7} = \frac{2923.531}{1416.331} = 2.064$$

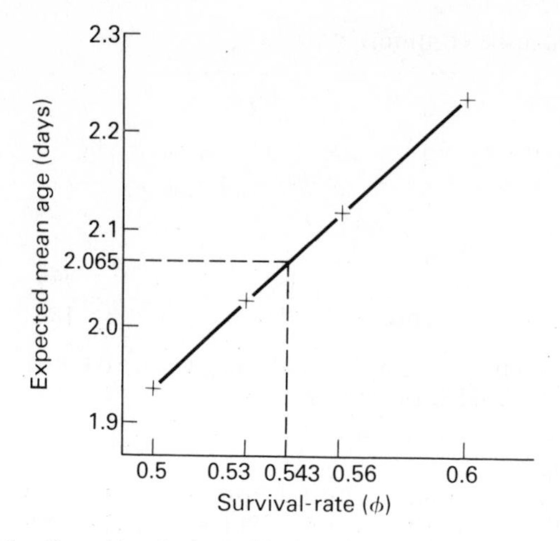

Fig. 2.1. Estimation of survival-rate (Jackson's negative method).

A ϕ-value of 0.543 can, therefore, be accepted, and N_7 estimated:

$$\hat{N}_7 = \frac{\hat{M}_7.(n_7+1)}{(m_7+1)} = \frac{1416.331 \times 1559}{170} = 12\ 989$$

Moreover, Bailey (1951) also gave the following formulae for estimating the standard errors of $\hat{\phi}$ and $\hat{N}_7$, where:

$$F_7 = \sum_j r_j \phi^{(7-j)}(7-j)^2$$

$$\mathrm{SE}_{\hat{N}_7} = \sqrt{\frac{\hat{N}_7^2}{n_7}\left(\frac{\hat{N}_7 F_7}{\hat{M}_7 F_7 - \hat{T}_7^2} - 1\right)} = 1871$$

$$\mathrm{SE}_{\hat{\phi}} = \sqrt{\frac{\hat{\phi}^2 \hat{N}_7 \hat{M}_7}{n_7(\hat{M}_7 F_7 - \hat{T}_7^2)}} = 0.033$$

The estimated size of the tsetse fly population was 12 989; its estimated survival rate per week was 0.543.

Jackson's negative method requires several visits to the study-site, but provides only one estimate of population size, and one of survival-rate – which is additionally assumed to remain constant throughout. Gain, at a variable rate, is allowed. The parallels between this and Jackson's positive method are obvious: it is an extension of the Petersen estimate which allows gain, but no loss, and estimates population size for the final sampling occasion; it provides little return, in terms of estimates, for the effort expended; but it is ready-made for situations in which marks *can* only be sought once.

In other words Jackson's negative method is useful when capturing and marking are relatively easy, but screening for marks is difficult or time-consuming. Small flies, for example, can be readily marked with ultra-violet sensitive dust. But the unequivocal classification of marked and unmarked individuals requires a lens, a darkened room, and an ultra-violet lamp: in short, a trip to the laboratory. Consequently, marks *can* often only be sought once: at the end of the study. The utility of Jackson's negative method is in strict proportion to the frequency with which such situations arise. As with the positive method, the situations are probably not common – but they occur sufficiently frequently for this to be a useful model in the capture-recapture repertoire.

2.8 The Fisher-Ford method

The method developed by Fisher and Ford (1947) is very similar to Jackson's negative method, but relies on several marking occasions and several recaptures. Population size is estimated from the modified Petersen estimate by assuming that the ratio of marks to total individuals in the day i sample is the same as in the total population:

$$\hat{N}_i = \frac{(n_i + 1)}{(m_i + 1)}.M_i$$

However, M_i – the marks at risk on day i – is itself unknown, and must be estimated before N_i can be calculated. (Note that we are considering marks, not marked individuals.)

The procedure for calculating $\hat{M}_i$ is fairly complicated and somewhat indirect, and involves comparing the days survived by marks in the samples with the days survived by marks in the population. These should, on average, be the same, but because of sampling error, this equality is unlikely to be exact on any particular day. It is better, therefore, to add the days survived from different sampling occasions together, and then compare the sum from the samples with the sum from the populations. By doing this, discrepancies should cancel each other out to some extent.

Consider the samples first. The total number of marks caught on day i is given by:

$$m_i = \sum_j m_{ij}$$

where j – the day on which the mark was given – varies from 1 to $i-1$. Each mark is $(i-j)$ days old, and the total age of all marks is, therefore:

$$\sum_j m_{ij}(i-j)$$

This is, in other words, the number of days survived by the marks caught in the day i sample.

If this quantity is summed over all days, we obtain the 'total days survived' by marks throughout the experiment:

$$\sum_i \sum_j m_{ij}(i-j)$$

Note that this is not an estimate, but the calculation of a quantity that we actually observe: 'observed total days survived' or 'observed TDS'.

Now consider the whole population, where A_i is the mean age of marks on day i. Immediately before the day i sample there are M_i marks at risk. During the sample some are removed and examined, but at the end they are replaced – along with r_i new marks. Immediately *after* the day i sample there are therefore $(M_i + r_i)$ marks at risk in the population. At this time the M_i marks have a total age of $A_i M_i$, and the r_i marks – released immediately before – a total age of zero. The mean age of all the marks is, therefore, given by:

$$\frac{A_i M_i + 0}{M_i + r_i}$$

One day later, immediately before the day $i+1$ sample, all these marks are one day older and consequently:

$$A_{i+1} = \frac{A_i M_i}{M_i + r_i} + 1$$

(Note that if some previously-marked individuals are damaged during the day i sample, M_i in the above equation will have to be altered accordingly.)

All r_i's are known. Assume initially that all M_i's are also known. A_2 – the mean age of marks one day after the first (and only previous) sample – is obviously *1*. A_3 can, therefore, be calculated. This in turn means that A_4 can be calculated, and so on throughout the sequence of A_i's. We now have a term, $A_i m_i$, which is an estimate of the days survived by the marks caught in the day i sample. Similarly, we can compute $\sum_i A_i m_i$, the estimated total days survived (estimated TDS), which should equal the observed TDS previously calculated from the samples.

In other words, a sequence of M_i's will lead to a sequence of A_i's, and finally to a value for $\sum_i A_i m_i$. We are attempting, however, to estimate the M_i's themselves. The estimates we need, therefore, are those that give a $\sum_i A_i m_i$ which *is* equal to the observed TDS.

Remember that immediately after the day i sample there are $(M_i + r_i)$ marks at risk. Yet, between day i and day $i+1$ some of these are lost; only $\phi(M_i + r_i)$ survive. (ϕ is the daily survival-rate, assumed to remain constant). In other words, immediately before the day $i+1$ sample there are $\phi(M_i + r_i)$ marks at risk, or:

$$M_{i+1} = \phi(M_i + r_i)$$

All r_i's are known. Assume initially that ϕ is also known. M_1 – the marks at risk before the first sample – is obviously zero. M_2 can, therefore, be calculated ($= \phi r_1$). Now r_2, ϕ and M_2 are all known, and M_3 can also be calculated, followed by all the other M_i's in sequence.

In other words, a particular value of ϕ has led, via a sequence of M_i's, to a particular value of $\sum_i A_i m_i$. Therefore, the final step is to plot a graph of ϕ against $\sum_i A_i m_i$ (four points, appropriately spaced, are sufficient). The ϕ-value that would lead to a $\sum_i A_i m_i$-value equal to the observed TDS can be read from this graph. This is our best estimate of ϕ, and it can be used to obtain our best sequence of $\hat{M}_i$'s. The $\hat{N}_i$'s can then be calculated.

Finally, the *number* of losses between days i and $i+1$ (L_i), and the number of gains between days i and $i+1$ (B_i) can be estimated from the following equations:

$$\hat{L}_i = (1 - \hat{\phi})\hat{N}_i$$

$$\hat{B}_i = \hat{N}_{i+1} - \hat{\phi}\hat{N}_i$$

Example. The following data are given by Dowdeswell, Fisher and Ford (1940) for the Common Blue Butterfly on Tean (Isles of Scilly). These data are arranged in a trellis (Table 2.2). There were, for instance, 15 day *5* marks caught on day *6*.

Table 2.2 Common Blue capture-recapture data (Dowdeswell *et al.*, 1940).

Day i	Captured n_i	Released r_i	Time of release of marks, j: *1*	*2*	*4*	*5*	*6*	*8*	*9*	*12*	*13*
			Recaptured marks, m_{ij}								
1	—	40									
2	43	40	5								
4	13	12	0	3							
5	52	50	3	8	5						
6	56	51	6	12	6	15					
8	52	52	4	10	3	16	14				
9	50	50	4	5	1	11	5	14			
12	15	15	1	1	1	3	1	5	5		
13	20	20	1	1	2	3	2	7	8	6	
14	7	—	0	0	0	2	2	4	1	0	4

Table 2.3 Observed days survived computed from Common Blue data.

Day, i	m_i	$\sum_j m_{ij}(i-j)$
1	0	0
2	5	5
4	3	6
5	16	41
6	39	105
8	47	176
9	40	145
12	17	91
13	30	152
14	13	67
		$788 = \sum_i \sum_j m_{ij}(i-j)$ (observed TDS)

The first step is to calculate the days survived by marks in the samples. Thus, for example on day 5:

$$\sum_j m_{5j}(5-j) = (3\times 4)+(8\times 3)+(5\times 1) = 41$$

Note immediately that captures were not made on all days, but that the term (i–J) still calculates the age of marks correctly. Other days are treated similarly (Table 2.3).

The next step is to find a survival-rate, ϕ, that will give an estimated TDS equal to 788. Quite arbitrarily, we try a survival-rate of 0.8 first:

$$M_1 = 0$$
$$M_2 = 0.8(0+40) = 32$$
$$M_3 = 0.8(32+40) = 57.6$$

Table 2.4 Estimated days survived computed from Common Blue data with a survival-rate (ϕ) of 0.8.

Day, i	M_i	A_i	$A_i m_i$
1	0	—	—
2	32.0	1	5
4	46.1	2.44	7.33
5	46.5	2.94	47.02
6	77.2	2.42	94.22
8	82.0	3.46	162.39
9	107.2	3.14	124.56
12	80.5	5.12	87.12
13	76.4	5.32	159.57
14	77.1	5.22	67.81
			755.02 (estimated TDS)

But no captures were made on day *3*, and we proceed immediately to day *4*:

$$M_4 = 0.8 \times 57.6 = 46.08$$

Similarly:

$$A_2 = 1$$

$$A_3 = \frac{1 \times 32}{32+40} + 1 = 1.44$$

and

$$A_4 = 1.44 + 1 = 2.44$$

Using these formulae, modified by the fact that samples were not taken on all days, Table 2.4 can be constructed.

Our ϕ-value is, therefore, too low.

If we try a ϕ of 0.9, we get an estimated TDS of 872.91 which is too high. Our best estimate of ϕ must, therefore, lie between 0.8 and 0.9. We can add two further points:

ϕ	$\Sigma A_i m_i$
0.75	698.04
0.85	813.62

and plot the resulting graph (Fig. 2.2). This suggests that a ϕ value of 0.828 would lead to a $\sum_i A_i m_i$ -value of 788 – which is checked in Table 2.5.

This check can obviously be viewed as successful. ϕ has, therefore, been estimated, and the required sequence of M_i's produced. This sequence can be used to estimate the N_i's, L_i's and B_i's (Table 2.6).

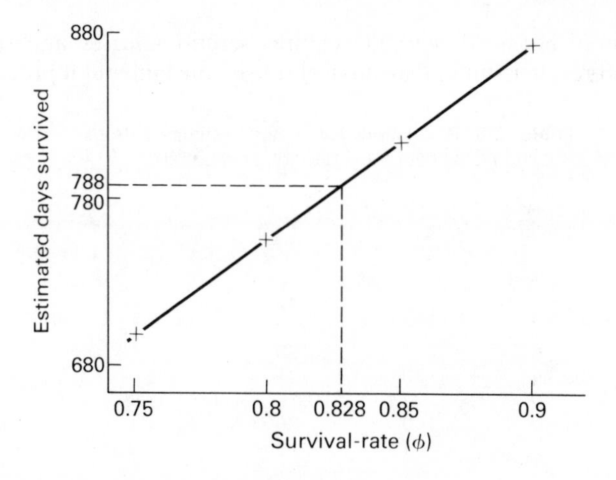

Fig. 2.2. Estimation of survival-rate (Fisher-Ford method).

Table 2.5 Estimated days survived computed from Common Blue data with a survival-rate (ϕ) of 0.828.

Day, i	M_i	A_i	$A_i m_i$
1	0	—	—
2	33.1	1	5
4	50.2	2.45	7.36
5	51.5	2.98	47.66
6	84.1	2.51	97.97
8	92.7	3.56	167.46
9	119.8	3.28	131.32
12	96.5	5.32	90.37
13	92.4	5.60	168.03
14	93.1	5.60	72.85
			788.02

Note that because of the days on which no data were collected, it is not possible to estimate B_i for all days.

The analysis is now essentially complete. It will be obvious, however, that the assumption of a constant survival-rate has been a central feature of the whole method. Tests of this assumption are described in Chapter 4.

The estimated size of the Common Blue Butterfly population fell fairly steadily between days 2 *and* 14 *from 243 to 53; the estimated daily survival-rate was* 0.828.

The Fisher-Ford method requires several releases and several recaptures. It assumes that survival-rate is constant, but it produces a

Table 2.6 Population parameters estimated from Common Blue data (Dowdeswell *et al.*, 1940).

Day, i	$\hat{N}_i$	$\hat{L}_i$	$\hat{B}_i$
1	—	—	—
2	243	42	—
4	176	30	15
5	161	28	−13
6	120	21	—
8	102	18	65
9	149	26	—
12	86	15	−8
13	63	11	1
14	53	9	—

full series of estimates for population size, gain and loss. It provides more information than any of the models described so far, and does so with at least as high an efficiency. The models to be described next, however – Jolly's and Manly and Parr's – do all that the Fisher-Ford does, but also allow survival-rate to vary. In comparison with these, Fisher-Ford method seems relatively *in*efficient.

There are, nevertheless, many studies involving marked release over an extended period which accumulate relatively little data – especially in terms of recaptures. The Fisher-Ford method has a major advantage in such cases. Its estimates are derived by grouping data together to produce a single survival-rate. Samples are combined, and sampling errors should, therefore, tend to cancel out. Overall sampling error should be minimized. The poorer the data the more useful this is. In comparison, the methods of Jolly and Manly and Parr deal with smaller subsets of the data, and are hence more prone to sampling errors. Such an argument is only valid, however, if the survival-rate is constant, or nearly so. If it is subject to considerable variation, then the minimization of sampling error is spurious: the various days' data are not samples of the same thing at all, and combining them will hide real and important differences.

Overall the Fisher-Ford method has undoubted advantages over simpler models, and is particularly useful when the data are relatively poor, and the survival-rate approximately constant. Its utility in multiple recapture studies will be discussed further in Chapter 3.

2.9 Jolly's stochastic method

Like the Fisher-Ford method, Jolly's stochastic method (1965) requires several marking occasions and several recaptures. When marked individuals are being considered, however, only the most recent mark is noted – all previous marks are ignored. Each marked individual in the day i sample contributes only one mark (its most recent) to the total, m_i.

Similarly, the number of 'marks' at risk in the day i population (M_i) means the number of 'marked individuals' at risk in this case.

The estimation of M_i is the central feature of Jolly's stochastic method, and is, consequently, the first step in the analysis. We start with the simple rearrangement:

$$M_i = m_i + (M_i - m_i)$$

m_i is known, and so M_i can be estimated if we can estimate $(M_i - m_i)$: the number of marked individuals at risk on day i that were *not* caught.

Of these $(M_i - m_i)$ individuals, some will be caught subsequently (on day $i+1$ or day $i+2$ or day $i+3$ etc.). They will be recognizable because their most recent mark will be one given to them on a day *prior* to day i. We may, therefore, define z_i as the number of individuals marked before day i, not caught on day i itself, but caught subsequently. In other words, of the $(M_i - m_i)$ individuals at risk on day i, z_i are caught subsequently.

Immediately after the day i sample, these $(M_i - m_i)$ individuals are joined by the r_i individuals released on day i with a day i mark. We can define y_i as the number of these r_i individuals that are caught again subsequently. If we assume that the fates of the two groups – r_i and $(M_i - m_i)$ – run parallel after day i, then it should be true that:

$$\frac{z_i}{M_i - m_i} = \frac{y_i}{r_i}$$

m_i, r_i, y_i and z_i can all be obtained directly from the data, and the M_i's can, therefore, be estimated:

$$\hat{M}_i = m_i + \frac{z_i r_i}{y_i}$$

Population size is then given, as usual, by a modified Petersen estimate:

$$\hat{N}_i = \frac{\hat{M}_i(n_i + 1)}{(m_i + 1)}$$

Immediately after the day i sample there are $(M_i - m_i + r_i)$ marked individuals in the population: the number at risk before the sample, minus those removed during it, plus the new marks added at the sample's end. Of these marked individuals there will, by definition, be M_{i+1} individuals that survive until the day $i+1$ sample. The survival-rate from day i until day $i+1$ (ϕ_i) can, therefore, be estimated by:

$$\hat{\phi}_i = \frac{\hat{M}_{i+1}}{\hat{M}_i - m_i + r_i}$$

Finally, the number of additions to the population between days i and $i+1$ is given by:

$$\hat{B}_i = \hat{N}_{i+1} - \hat{\phi}_i \hat{N}_i$$

Example. The data in Table 2.7 are those used by Jolly (1965) himself, for an apple orchard population of female black-kneed capsids. These were caught on thirteen occasions at, alternately, three- and four-day intervals. For present purposes the study will be assumed to have lasted for just thirteen days.

Remember that only the most recent marks have been considered – the day 6 sample, for instance, may in fact have been of

Table 2.7 Capture-recapture data from Jolly (1965).

Day	Captured	Released	Time of release of marks, j											
i	n_i	r_i	1	2	3	4	5	6	7	8	9	10	11	12
			Recaptured marks, m_{ij}											
1	—	54												
2	146	143	10											
3	169	164	3	34										
4	209	202	5	18	33									
5	220	214	2	8	13	30								
6	209	207	2	4	8	20	43							
7	250	243	1	6	5	10	34	56						
8	176	175	0	4	0	3	14	19	46					
9	172	169	0	2	4	2	11	12	28	51				
10	127	126	0	0	1	2	3	5	17	22	34			
11	123	120	1	2	3	1	0	4	8	12	16	30		
12	120	120	0	1	3	1	1	2	7	4	11	16	26	
13	142	—	0	1	0	2	3	3	2	10	9	12	18	35

the following form: 'captured 209; 2 with a day *1* mark, 2 with day *1*, *2* and *3* marks, 3 with day *1* and *5* marks, 4 with a day *2* mark, 5 with day *2*, *4* and *5* marks, 6 with a day *3* mark, 10 with day *3* and *4* marks, 10 with day *3*, *4* and *5* marks, 10 with a day *4* mark, 5 with day *4* and *5* marks, 20 with a day *5* mark; 207 released'. This reduces to row *6* of the trellis.

The m_i's, y_i's and z_i's must now be calculated.

m_i – the total number of marked individuals captured on day i – is the sum of the ith row of the trellis. For instance:

$$m_6 = 2+4+8+20+43 = 77$$

y_i – the total number of the r_i individuals that are caught subsequently – is the sum of the ith column of the trellis. For instance:

$$y_6 = 56+19+12+5+4+2+3 = 101$$

z_i – the total number of individuals marked before day i, not caught on day i, but caught subsequently – is slightly more difficult to calculate. It is the sum of those m_{ij}'s which are both in columns to the left of column i, and in rows below row i. For instance, z_6 is the sum of the m_{ij}'s in the 5×7 rectangle comprising those parts of columns *1* to *5* that are also in rows *7* to *13*:

$$z_6 = 1+6+5+10+34+0\ldots+1+0+1+0+2+3 = 121$$

Table 2.8 can, therefore, be drawn up.

Note that these figures have been tabulated either directly from the data (in the case of the r_i's), or by simple addition of specific combinations of m_{ij}'s.

Table 2.8 Preliminary computations for Jolly's method.

Day, i	r_i	m_i	y_i	z_i
1	54	—	24	—
2	143	10	80	14
3	164	37	70	57
4	202	56	71	71
5	214	53	109	89
6	207	77	101	121
7	243	112	108	110
8	175	86	99	132
9	169	110	70	121
10	126	84	58	107
11	120	77	44	88
12	120	72	35	60
13	—	95	—	—

Calculation of the $\hat{M}_i$'s, and thus the $\hat{N}_i$'s, $\hat{\phi}_i$'s and $\hat{B}_i$'s, is now simply achieved by substituting into the appropriate equations. For instance:

$$\hat{M}_6 = 77 + \frac{121 \times 207}{101} = 324.99$$

$$\hat{M}_7 = 112 + \frac{110 \times 243}{108} = 359.50$$

$$\hat{N}_6 = \frac{324.99 \times 210}{78} = 874.97$$

$$\hat{N}_7 = \frac{359.5 \times 251}{113} = 798.54$$

$$\hat{\phi}_6 = \frac{359.5}{324.99 - 77 + 207} = 0.790$$

$$\hat{B}_6 = 798.54 - 0.790 \times 874.97 = 107.31$$

Moreover, Jolly (1965) gave the following formulae for standard errors:

$$\mathrm{SE}_{\hat{N}_i} = \sqrt{\hat{N}_i(\hat{N}_i - n_i)\left\{\frac{\hat{M}_i - m_i + r_i}{\hat{M}_i}\left(\frac{1}{y_i} - \frac{1}{r_i}\right) + \frac{1}{m_i} - \frac{1}{n_i}\right\}}$$

$$\mathrm{SE}_{\phi_i} = \hat{\phi}_i\left\{\frac{(\hat{M}_{i+1} - m_{i+1})(\hat{M}_{i+1} - m_{i+1} + r_{i+1})}{\hat{M}_{i+1}^2}\left(\frac{1}{y_{i+1}} - \frac{1}{r_{i+1}}\right)\right.$$

$$\left. + \frac{\hat{M}_i - m_i}{\hat{M}_i - m_i + r_i}\left(\frac{1}{y_i} - \frac{1}{r_i}\right)\right\}^{\frac{1}{2}}$$

$$\mathrm{SE}_{B_i} = \left\{ \frac{\hat{B}_i^2(\hat{M}_{i+1} - m_{i+1})(\hat{M}_{i+1} - m_{i+1} + r_{i+1})}{\hat{M}_{i+1}} \left(\frac{1}{y_{i+1}} - \frac{1}{r_{i+1}} \right) \right.$$

$$+ \frac{\hat{M}_i - m_i}{\hat{M}_i - m_i + r_i} \left(\hat{\phi}_i r_i \left[\frac{n_i}{m_i} - 1 \right] \right)^2 \left(\frac{1}{y_i} - \frac{1}{r_i} \right)$$

$$+ \frac{(\hat{N}_i - n_i)(\hat{N}_{i+1} - B_i)(1 - \hat{\phi}_i)\left(1 - \frac{m_i}{n_i}\right)}{\hat{M}_i - m_i + r_i}$$

$$\left. + \frac{\hat{N}_{i+1}(\hat{N}_{i+1} - n_{i+1})\left(1 - \frac{m_{i+1}}{n_{i+1}}\right)}{m_{i+1}} + \frac{\hat{\phi}_i^2 \hat{N}_i(\hat{N}_i - n_i)\left(1 - \frac{m_i}{n_i}\right)}{m_i} \right\}^{\frac{1}{2}}$$

Thus, overall, the results in Table 2.9 can be drawn up.

Table 2.9 Results computed from Jolly's (1965) data.

Day, i	$\hat{M}_i$	$\hat{N}_i$	$\mathrm{SE}_{\hat{N}_i}$	0_i	SE_{0_i}	$\hat{B}_i$	$\mathrm{SE}_{\hat{B}_i}$
1	0	—	—	0.649	.093	—	—
2	35.02	468	151	1.015	.110	288	179
3	170.54	763	129	0.867	.105	289	138
4	258.00	951	140	0.564	.059	396	120
5	227.73	932	124	0.836	.073	96	111
6	324.99	875	94	0.790	.068	108	75
7	359.50	799	72	0.651	.052	130	56
8	319.33	650	59	0.985	.093	–13	52
9	402.13	627	59	0.686	.077	47	34
10	316.45	477	49	0.884	.118	82	40
11	317.00	504	64	0.771	.126	71	41
12	277.71	460	68	—	—	—	—

Jolly's stochastic method is a considerable improvement on any of the models so far described. It shares with the Fisher-Ford method the advantage of estimating parameters over an extended period, but also allows survival-rate to vary from day to day. Related to this, survival is probablistic or stochastic, rather than deterministic – an important gain in realism.

The advantages of this method are obvious. The difficulty is in deciding when *not* to use it. Of course, if time is limited, or marking or screening restricted, then a simpler model must be used. Moreover, if the data are extensive but somewhat scanty, then the Fisher-Ford method may quite possibly be the best. Finally, if the data are full and extensive, and multiple recaptures are very common, then the Manly-Parr method – to be described next – has considerable claims of its own. The choice of a model for multiple capture-recapture data will be discussed in the next chapter.

Generally, however, there will be many situations where the degree of sophistication shown by Jolly's stochastic method makes it the most appropriate model.

2.10 The Manly-Parr method

This method, proposed by Manly and Parr (1968), is unusual in two respects. The first is the way in which the data are tabulated. The method can only be made fully clear, therefore, by reference to the *Example.* The second is the dependence of all other estimates on an estimate of p_i: the proportion of the population that are caught on day *i.*

It is convenient, initially, to assume that p_i has been estimated. The population size is then estimated by:

$$\hat{N}_i = \frac{n_i}{\hat{p}_i}$$

Furthermore, if we let a_i be the number of individuals captured on both day *i* and day $i+1$; and let ϕ_i be the survival rate from day *i* until day $i+1$, then:

$$a_i = r_i \phi_i p_{i+1}$$

i.e. of the r_i individuals released on day *i*, a proportion, ϕ_i, survive until day $i+1$ when a proportion, p_{i+1}, are captured.
Therefore:

$$\hat{\phi}_i = \frac{a_i}{r_i \hat{p}_{i+1}}$$

Finally, B_i – the number born between days *i* and $i+1$ – is estimated by:

$$\hat{B}_i = \hat{N}_{i+1} - \hat{\phi}_i \hat{N}_i$$

The estimation of the p_i's will, therefore, enable the $\hat{N}_i$'s, $\hat{\phi}_i$'s and $\hat{B}_i$'s to be calculated.

On day *i* there are N_i animals alive, which can be divided into four mutually exclusive groups, sizes W_i, X_i, Y_i and Z_i:

	Captured on day i	Not captured on day i
Captured at least once before day i and once after day i	W_i	X_i
Not captured before day i and/or not captured after day i	Y_i	Z_i
Total	n_i	$N_i - n_i$

In other words, the classification by rows divides the day i population into two groups, and all individuals are members of one group or the other, but not both. Within each group the ratio of captured to not-captured should be approximately equal, and should, therefore, approximate this same ratio in the total population:

$$\frac{W_i}{X_i} \simeq \frac{Y_i}{Z_i} \simeq \frac{n_i}{N_i - n_i}$$

This amounts to saying:

$$\frac{W_i}{X_i} \text{ estimates } \frac{n_i}{N_i - n_i}$$

But we need an estimate of $p_i \left(= \frac{n_i}{N_i} \right)$. This is therefore given by:

$$\hat{p}_i = \frac{W_i}{W_i + X_i}$$

The calculations of W_i and X_i are best explained within the *Example.*

Example. The following (hypothetical) data are actually somewhat inadequate for this method (see below, p. 42), but will provide a manageable illustration.

Day *1*: 10 released (red)
Day *2*: 12 captured; 5 red; 12 released (blue)
Day *3*: 10 captured; 2 red, 2 red-blue, 3 blue; 10 released (green)
Day *4*: 8 captured; 1 red-blue, 1 red-green, 2 blue-green, 2 green; 8 released (brown)
Day *5*: 10 captured; 1 red-blue, 2 blue, 1 blue-green-brown, 2 green-brown, 1 brown.

The tabulation of these data lies at the heart of the Manly-Parr method. Initially this tabulation may seem strange. 'Days' should be entered along the top of the table, and for day *1* ten *1*'s (the number released) should be entered vertically (Table 2.10a).

Next consider day *2*. Five individuals bore a red mark, and should be entered (as *1*'s) opposite the first five day *1* captures. Seven individuals were captured for the first time, and should be entered (as *1*'s) in lines 11–17 (Table 2.10b).

On day *3*, two individuals were caught with a red mark only: individuals which were caught on day *1* but not on day *2*. Consequently, *1*'s should be entered in column 3 lines 6 and 7; while in column 2 lines 6 and 7, *0*'s should be entered to signify that these individuals were alive but not captured.

Two red-blue individuals were also captured. *1*'s are, therefore, entered in lines 1 and 2 of column 3; while in column 2 lines 1 and 2 the *1*'s (because they have *1*'s either side of them) are converted to +'s. The three blue and three unmarked individuals are entered as before, and after day *3* the data appear as in Table 2.10c.

The same procedure is then applied for day *4*: *1*'s are entered as appropriate; *1*'s with *1*'s either side of them are converted to +'s; blanks with *1*'s either side of them are converted to *0*'s. The day *4* data appear as in Table 2.10d.

Finally, after day *5*, the data appear as in Table 2.10e.

The table is now complete, and the reason for this tabulation should be apparent: *1*'s are used because they can be converted to +'s; the number of +'s on day i is W_i: the number of individuals caught on day i and also caught both before and after day i; and the number of *0*'s on day i is X_i: the number of individuals not caught on day i, but known to be alive then because they were caught at least once before and once after day i. p_i can, therefore, be estimated:

$$\hat{p}_2 = \frac{4}{4+2} = 0.67$$

$$\hat{p}_3 = \frac{3}{3+4} = 0.43$$

$$\hat{p}_4 = \frac{3}{3+3} = 0.5$$

And thus, for example:

$$\hat{N}_3 = \frac{n_3}{\hat{p}_3} = \frac{10}{0.43} = 23.3$$

$$\hat{\phi}_3 = \frac{a_3}{r_3 \cdot \hat{p}_4} = \frac{5}{10 \times 0.5} = 1.00$$

and overall:

Day (i)	$\hat{N}_i$	$\hat{\phi}_i$	$\hat{B}_i$
1	—	—	—
2	18.0	0.97	5.8
3	23.3	1.00	−7.3
4	16.0	—	—
5	—	—	—

Table 2.10 Tabulation of data for Manly-Parr method. a) after day *1*, b) after day *2*, c) after day *3*, d) after day *4*, e) after day *5*.

	(a) Day:	(b) Day:		(c) Day:			(d) Day:				(e) Day:				
	1	*1*	*2*	*1*	*2*	*3*	*1*	*2*	*3*	*4*	*1*	*2*	*3*	*4*	*5*
(1)	*1*	*1*	*1*	*1*	+	*1*	*1*	+	*1*		*1*	+	*1*		
(2)	*1*	*1*	*1*	*1*	+	*1*	*1*	+	*1*		*1*	+	*1*		
(3)	*1*	*1*	*1*	*1*	*1*		*1*	+	*0*	*1*	*1*	+	*0*	*1*	
(4)	*1*	*1*	*1*	*1*	*1*		*1*	*1*			*1*	+	*0*	*0*	*1*
(5)	*1*	*1*	*1*	*1*	*1*		*1*	*1*			*1*	*1*			
(6)	*1*	*1*		*1*	*0*	*1*	*1*	*0*	+	*1*	*1*	*0*	+	*1*	
(7)	*1*	*1*		*1*	*0*	*1*	*1*	*0*	*1*		*1*	*0*	*1*		
(8)	*1*	*1*		*1*			*1*				*1*				
(9)	*1*	*1*		*1*			*1*				*1*				
(10)	*1*	*1*		*1*			*1*				*1*				
(11)			*1*		*1*	*1*		*1*	+	*1*		*1*	+	+	*1*
(12)			*1*		*1*	*1*		*1*	+	*1*		*1*	+	*1*	
(13)			*1*		*1*	*1*		*1*	*1*			*1*	*1*		
(14)			*1*		*1*			*1*				*1*	*0*	*0*	*1*
(15)			*1*		*1*			*1*				*1*	*0*	*0*	*1*
(16)			*1*		*1*			*1*				*1*			
(17)			*1*		*1*			*1*				*1*			
(18)						*1*			*1*	*1*			*1*	+	*1*
(19)						*1*			*1*	*1*			*1*	+	*1*
(20)						*1*			*1*				*1*		
(21)										*1*				*1*	*1*
(22)										*1*				*1*	
(23)															*1*

Despite its apparent novelty, the Manly-Parr method is, in fact, closely related to the other capture-recapture models – particularly Jolly's. W_i is the proportion (k) of the marks caught on day i (m_i) that had been caught before, and are going to be caught again. X_i is the *same* proportion (we assume) of the $M_i - m_i$ marks at risk on day i that were *not* caught. This means that:

$$\hat{p}_i = \frac{W_i}{W_i + X_i} = \frac{km_i}{km_i + k(M_i - m_i)}$$

There are two points to note from this. Firstly that it reduces to:

$$\hat{p}_i = \frac{m_i}{M_i}$$

and that the estimation of N_i reduces to a Petersen estimate. And secondly, that the estimation of $M_i - m_i$ (actually $k(M_i - m_i)$ in this case) is a central feature of the model – just as it was in Jolly's method. The two models estimate this quantity in totally different ways.

The real novelty of the Manly-Parr method is that it makes one less assumption than the other models. Even Jolly's method assumed that survival was independent of age; that the daily survival-rate of individuals was the same irrespective of when they were last caught. But at no time in the Manly-Parr method is an age-independent survival-rate assumed. This is the only advantage it has over Jolly's method, but it is an extremely important one. Survival is probably never strictly independent of age, and is sometimes very far from independent.

This disadvantage of the Manly-Parr method is that large numbers of multiple-recaptures are necessary before the estimates can be relied on. Manly and Parr (1968) themselves state that W_i should at least exceed ten. This is because the method is based on a comparatively restricted class of individuals. Ironically, it is this same reliance which releases the method from the assumption of age-independent survival. Once again, a single facet of the method is both a major strength and a major weakness.

The relative utilities of Manly-Parr, Fisher-Ford and Jolly's methods are discussed further in the next chapter.

2.11 The partitioning of loss and gain

Finally, one of the major problems with all capture-recapture models is that death and emigration are combined as loss, and birth and immigration combined as gain. It is possible, however, to repartition loss and gain, and estimate birth, death, immigration and emigration separately. The method for doing this was initially developed by Jackson (1939) in the context of his positive and negative

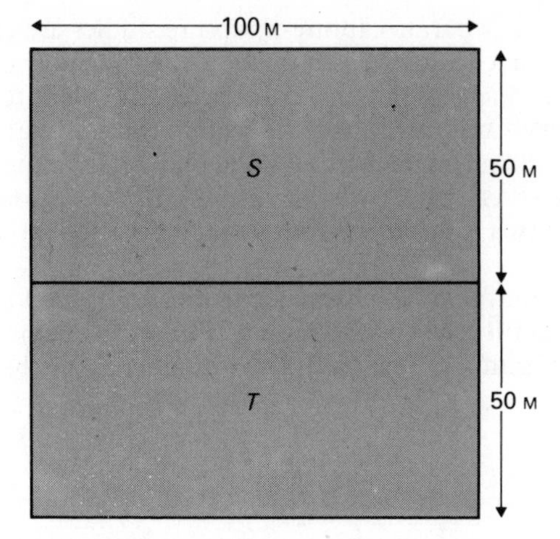

Fig. 2.3 Study-site for the partitioning of loss and gain.

methods, but it can be applied to any model in which gain and loss are estimated.

The method deals with numbers rather than rates, and these must first be computed. In general, the number of additions to a population between days i and $i+1$ is given by:

$$\hat{B}_i = \hat{N}_{i+1} - \hat{\phi}_i \hat{N}_i$$

Note that in those cases where the gain-*rate* has been computed:

$$\hat{B}_i = \hat{b}_i \hat{N}_{i+1}$$

i.e. b_i is the proportion of the day $i+1$ population added between days i and $i+1$.

The number of losses between days i and $i+1$ is given by:

$$\hat{L}_i = (1 - \hat{\phi}_i)\hat{N}_i$$

Imagine a population of uniform density, of which we are considering a part. The basis of Jackson's method is that the numbers of births and deaths in the sub-population are proportional to the area (volume) which it occupies. (With a uniform density, an increase in area means a proportional increase in population size, which means a proportional increase in the numbers of births and deaths.) The numbers of immigrations and emigrations, on the other hand, are proportional to the length of the sub-population's perimeter: the longer the boundary, the more likely it is that individuals will pass across it.

It is convenient, initially, to concentrate on gains. Imagine a study area of the type shown in Fig. 2.3.

Imagine also that separate capture-recapture studies are carried out simultaneously in each of the two small rectangles. Individuals must be given marks which are not only date-specific but also area-specific. In practical terms this means that we require twice as many types of mark as we would normally need. $\hat{B}_i$'s can then be computed for both S and T. However, by classifying marks as day *1*, day *2*, etc., irrespective of area, $\hat{B}_i$'s can also be computed for the larger, square population.

On day i, $\hat{B}_S$ and $\hat{B}_T$ are the estimated numbers of gains for the small populations, and $\hat{B}_R$ the corresponding estimate for the large. If α is the number of immigrations and β the number of births, then, by definition:

$$B_R = \alpha_R + \beta_R$$
$$B_S = \alpha_S + \beta_S$$

and

$$B_T = \alpha_T + \beta_T$$

Also, because births are proportional to area:

$$\beta_R = \beta_S + \beta_T$$

Immigration, however, is proportional to perimeter-length. The small areas each have perimeters of 300 m, and the large area one of 400 m. Thus:

$$\frac{\alpha_R}{400} = \frac{(\alpha_S + \alpha_T)}{600}$$

or

$$1.5\alpha_R = \alpha_S + \alpha_T$$

This means that:

$$1.5\alpha_R + \beta_R = \alpha_S + \alpha_T + \beta_S + \beta_T$$
$$B_R + 0.5\alpha_R = B_S + B_T$$
$$\alpha_R = 2(B_S + B_T - B_R)$$
$$\beta_R = B_R - \alpha_R$$

B_R, B_S and B_T have all been estimated. $\hat{\alpha}_R$ and $\hat{\beta}_R$ can, therefore, be calculated.

Similarly, if γ_R refers to the number of emigrations on day i and δ_R to the number of deaths, then:

$$\hat{\gamma}_R = 2(\hat{L}_S + \hat{L}_T - \hat{L}_R)$$

and

$$\hat{\delta}_R = \hat{L}_R - \hat{\gamma}_R$$

Jackson himself considered the situation in which a large, square population is sub-divided into four smaller, square populations. The appropriate equations in this case are:

$$\hat{\alpha}_R = \hat{B}_S + \hat{B}_T + \hat{B}_U + \hat{B}_V - \hat{B}_R$$
$$\gamma_R = \hat{L}_S + \hat{L}_T + \hat{L}_U + \hat{L}_V - \hat{L}_R$$
$$\hat{\beta}_R = \hat{B}_R - \hat{\alpha}_R$$
$$\hat{\delta}_R = \hat{L}_R - \hat{\gamma}_R$$

3 Interpretation

When investigating animal abundance, estimation plays an essential role, but must always be combined with another major facet: interpretation. This is a point which applies throughout biological research. All investigations should be prompted by a desire to answer a specific biological question. Yet the models and tests which are used are *mathematical* tools providing *mathematical* answers, with complete indifference to the biological context. It is the responsibility of the investigator to combine mathematics with biology, and provide a biological answer.

There are, of course, many aspects of interpretation which are only pertinent in context. Apparent fluctuations or trends can only be interpreted in the light of knowledge of the population concerned. Nevertheless, there are general comments which can be made. The first of these concerns unexpected results. Suppose, for example, that a particular trend is expected, and is being quantified. The interpretation is comparatively straightforward. But if the same trend appears unexpectedly, the situation is quite different. The 'status' of the trend has changed. Its unexpectedness demands a cautious interpretation because, by implication at least, there was no prior biological suspicion – and, therefore, no totally satisfactory biological explanation. The situation is analogous to the use of *a priori* and *a posteriori* tests, described in most statistics textbooks.

The conventional first step in interpretation is the calculation of a standard error, because it measures the accuracy or precision of an estimate. It ought to be the case, for instance, that the true value has a 95% probability of lying within 1.96 standard errors of the estimate (the figures *estimate – 1.96 SE* and *estimate + 1.96 SE* are called the '95% confidence limits' of the estimate). The smaller the standard error (the tighter the confidence limits), the more accurate is the estimate. In other words, small standard errors indicate a high degree of precision in the results; or, we might think, a high degree of reliability. But this connection between standard errors on the one hand, and accuracy, precision and reliability on the other, rests squarely on the presumption that the underlying model is valid. In reality the situation is often more complicated. There may, for instance, be a bias in the estimates: a consistent tendency to underestimate (or overestimate) the true values. Such a bias may occasionally be due to the way in which the model's formulae were derived. But more commonly the cause will be found in a discrepancy between the behaviour of the real world, and the way it is assumed to behave by the model. These discrepancies may also invalidate the standard error formula itself. Thus, reliability and accuracy have not only a statistical component, but also a biological one. It will be useful to reserve the term 'precision' for the quantity measured by standard errors in ideal situations. 'Reliability' and 'accuracy' can then be given a broader meaning, covering bias, precision itself, and also the biological aspects of accuracy. It will be possible, then, for results to be precise (because of the large amounts of data involved) but unreliable (because of the invalidity of the model employed).

The interpretation of capture-recapture results is, therefore, a process of several stages. First, the accuracy required of the results should be determined. Then, the results can be examined from a purely statistical point of view to see if such accuracy has been achieved. If it has not, we must return to the first stage to ask whether less accurate results would be better than no results at all. But if it has, we can move on to the third stage to ask whether precision is a good indication of overall reliability. This is done by assessing the validity of the underlying assumptions.

The rationale behind this process should be obvious. We ask first whether the data *would* be adequate if the population *were* ideal. Only if they would, is it worth considering the biological reliability. If the data are inadequate for an ideal population, then they will certainly be inadequate for a real one, and the biological aspects are irrelevant.

The first two of the three stages will now be examined in more detail. The third – testing the assumptions – is the subject matter of the next chapter.

Robson and Regier (1964) have recommended an accuracy of 0.5 (true N in the range $0.5N - 1.5N$) for preliminary studies or management surveys where only a rough idea of population size is needed; one of 0.25 (N in the range $0.75N$–$1.25N$) for more accurate management work; and one of 0.1 for careful work into population dynamics. Such recommendations form a useful guide-line. But, as was pointed out in Chapter 1, the required accuracy must be determined ultimately by each investigator, in response to the specific and unique problems which he is attempting to solve. Robson and Regier's *recommendations* must not be used by investigators as a means of avoiding personal responsibility.

The second, statistical stage of interpretation is, in reality, more complicated than it has been made to appear. Firstly, for the Fisher-

Ford method, there are no formulae for deriving standard errors. Moreover, even when they are available, the use of standard errors in capture-recapture has been questioned. Both Manly (1971) and Roff (1973a), with the help of their computer simulations, concluded (a) that many of the standard error formulae were themselves questionable; and (b) that estimates and standard errors were highly correlated, so that underestimates appeared more accurate than they really were, and overestimates less so. Consequently, this second stage can never consist simply of a comparison of confidence limits and accuracy requirements. Instead, there are several alternatives.

The first is to conclude that the imperfections of error-estimation make capture-recapture methods worthless, except in very exceptional circumstances. In view of the lack of a method which is *less* imperfect, this seems an unnecessarily pessimistic and restrictive conclusion. Nevertheless, all other alternatives should be prefaced by an acceptance of the fact that, without reliable error-estimates, extreme caution is essential. The second alternative, therefore, is simply to use standard errors when they are available, but to bear their imperfections very firmly in mind during the later stages of interpretation.

The third alternative is to look at the problem from what is essentially the opposite point of view. The object of this second stage is to establish whether the results would be adequate if the population were ideal. It would be instructive, therefore, to examine populations which are known to *be* ideal, and for which the true values of population parameters are known. The amounts of data leading to adequate results in ideal populations could then be determined.

One way of doing this is by simulating ideal populations in a computer, and subjecting them to an imaginary sampling programme. The results obtained can then be compared with the true values; and the procedure repeated several times, so that general properties (rather than specific investigations) are studied. In particular, the sampling intensity needed to obtain estimates, with a required degree of accuracy, from a given population can be determined. The second stage of interpretation becomes a comparison between the apparent sampling intensity of the population under investigation, and the sampling intensity required for an equivalent computer population. When sampling in the field is not sufficiently intense, we must return to the first stage. (If there are insufficient data for an ideal population, there will certainly be insufficient data for a real one.) When sampling is adequate, however, we can move on to consider just *how* ideal the population was (Chapter 4).

Roff (1973a) considered a closed population of 500 individuals. He found that the Petersen estimate required sampling intensities of

Table 3.1 Sampling intensities (as percentages) required to provide stated accuracies in various simulated populations (from Bishop and Sheppard, 1973).

Population size	Survival rate	Jolly		Accuracy			Fisher-Ford
		.1	.25	.5	.1	.25	.5
200	0.5			>12			>12
200	0.9			>12		>12	9
1000	0.5			>12		>12	9
1000	0.9	12	9	>5	12	9	5
3000	0.5	>12	12	9	12	12	9
3000	0.9	12	5		12	5	

around 55%, 28% and 15% for accuracies of 0.1, 0.25 and 0.5 respectively; and that Jolly's method, carried out over five days, required corresponding intensities of 40–50%, 25–30% and around 15%.

Bishop and Sheppard (1973) considered the methods of Jolly and Fisher-Ford in a variety of simulated situations with sampling intensities of 5%, 9% and 12%. Studies lasted for either 10 or 20 'days', and the results examined were the *means* of each study – the numbers being kept constant during each simulation. Their results are summarized in Table 3.1. Note that when population size, survival-rate and/or sampling intensity are low, Fisher-Ford fares better than Jolly. Note, too, that further results obtained by Bishop and Sheppard suggest, if anything, that survival-rates are less reliable than population sizes at corresponding sampling intensities.

Table 3.2 Sample sizes required when *n*'s and *M*'s are approximately equal–see text (from Robson and Regier, 1964).

	Accuracy		
N	*0.5*	*0.25*	*0.1*
25	10	15	20
50	20	25	35
100	35	45	65
150	40	60	90
200	50	75	120
400	80	115	210
600	100	150	280
800	120	185	335
1000	140	210	380
2000	200	320	600
5000	330	530	1200
10 000	460	800	1600

Table 3.3 Sample sizes required when M_i is 1.5 n_i–see text (from Robson and Regier, 1964).

	Accuracy		
N	0.5	0.25	0.1
25	10	15	15
50	15	20	30
100	25	35	50
150	35	50	70
200	40	60	90
400	60	95	150
600	85	125	205
800	100	150	255
1000	110	165	300
2000	160	240	500
5000	260	430	900
10 000	380	630	1300

Fortunately, the properties of ideal populations can also be studied analytically. In particular, Robson and Regier (1964) considered the application of the Petersen estimate to an ideal population in which all assumptions hold. For a given population size, they calculated the pairs of sample sizes which would provide a particular level of accuracy with a particular probability (chosen, following statistical convention, to be 95%).

To simplify matters, let us assume that the first and second samples are the same size. Table 3.2 can then be drawn up. Suppose, for example, that we wish to estimate (with a 95% probability of being right) the size of a population which appears to be around 400. For an accuracy of 0.25 (95% probability that the population is in the range 300–500) we require samples of 115 individuals on both days. To increase the accuracy to 0.1 the samples would have to be of 210 individuals, and so on. The second stage of interpreting a Petersen estimate will now involve simply comparing N with the sample sizes used to derive it. Of course, if N can be guessed beforehand, sample sizes can be chosen to give the required accuracy.

Robson and Regier only considered the Petersen estimate. Yet the other models are all derived from it, usually with the 'day *1* sample' replaced by the 'marks at risk'. It would seem sensible, therefore, to utilize Table 3.2, at least as a rule of thumb, in those methods where n's and M's are of approximately equal size: Jackson's positive and negative methods. In the weighted mean, Fisher-Ford and Jolly's methods, however, marks tend to accumulate so that M_i exceeds n_i. Here, Table 3.3 – computed on the basis of M_i being 1.5 n_i – may be useful. Conversely, in the triple catch and Manly-Parr methods, M_i (or its equivalent) tends to be smaller than n_i. In these cases, Table

Table 3.4 Sample sizes required when n_i is 1.5 M_i–see text (from Robson and Regier, 1964).

	Accuracy		
N	0.5	0.25	0.1
25	15	20	20
50	25	30	45
100	40	55	75
150	50	75	100
200	60	90	135
400	95	140	220
600	125	185	310
800	150	220	380
1000	170	245	450
2000	240	360	750
5000	400	640	1400
10000	570	940	2000

3.4 – computed on the basis of n_i being 1.5 M_i – may be useful.

The suggestion that Tables 3.2, 3.3 and 3.4 should be applied to all capture-recapture models obviously involves some major simplifications. We have already seen, for instance, that when survival-rate is constant and sampling intensity fairly low, the Fisher-Ford method is more accurate than Jolly's. Also, the exact relationship between M_i and n_i will depend on the survival-rate. The higher this is, the more marks will accumulate, and the smaller the required sample size will be. In fact, many studies will involve survival-rates high enough to justify samples smaller than those in Tables 3.2, 3.3 and 3.4 (Further details are given by Robson and Regier (1964) and Seber (1973)). Nevertheless, the combination of these tables, the results of simulations, and a large amount of common sense, does offer a means of assessing the adequacy of results from a purely statistical point of view. This must, of course, remain a mere preliminary to the assessment of their biological adequacy (Chapter 4).

A rather more specific aspect of interpretation concerns results which are biologically impossible: estimates of survival-rate greater than 1, and negative values for $\hat{L}_i$ and $\hat{B}_i$. The reason for getting these is easy to see. $\hat{\phi}_i$, $\hat{B}_i$ and $\hat{L}_i$ *are* all estimates; they are all subject to error. Biologically they are confined, by definition, within certain limits. But statistically these limits can be breached; the models can, with imperfect data, produce 'impossible' answers. The correct interpretation of a $\hat{\phi}$-value greater than 1 would be that it was a combination of the true survival-rate (which may indeed be close to 1) and a positive error term. Our 'best' estimate of ϕ would, therefore, be 1.0. Similarly, if either $\hat{B}_i$ or $\hat{L}_i$ are negative, the interpretation must be that there is a negative error term; and our best estimate is zero. (A temptation which must be avoided is to interpret negative 'gains' as losses, and *vice versa.*)

The major concern of capture-recapture studies is population size. This implies that the term 'population' is fully understood. For animals living in habitat islands – ponds surrounded by land, small woods surrounded by fields, and so on – the limits of the population are easy to recognize. In such cases the whole population can be sampled, and capture-recapture methods do indeed measure 'population size'. But, more often than not, the experimental animals live in an extensive habitat, and population limits are defined artificially – usually for the investigator's convenience. In these cases the figure emanating from the theoretical model, and termed 'population size', requires further interpretation.

The usual procedure is to utilize the area of the artificially-defined population to convert population size to absolute density (numbers per unit area). If the limits, though artificial, *are* well-defined, this will frequently be justified. Often, however, the population limits in an extensive habitat will not only be artificial but also ill-defined. This will occur, in fact, whenever capture depends on the activity of the animals themselves (Chapter 5). The area occupied by the population will not be known exactly, and the interpretation of 'population size' will be even more than usually tentative. Minimizing the artificiality of the limits (and therefore minimizing the emigration-rate) will be useful; but knowledge of the animals' mobility in the presence of traps will be essential. Only then can the area from which animals are drawn be known, and population size converted to absolute density.

It will often be desirable to carry interpretation a stage further, by extrapolating from the area under investigation to the habitat as a whole. Needless to say, this too must be done cautiously. No ecologist could doubt that the distribution of animals *cannot* be discerned by the application of uninformed rationality. If absolute density in the experimental sub-population is to be considered typical of the population as a whole, then this consideration should be justified empirically. As with other aspects of interpretation, the connection between 'population size' and a meaningful biological answer is the responsibility of the investigator, and should be founded on his ecological expertise.

Finally, a topic left unfinished in Chapter 2 must be considered. The choice of which model to use is often not a matter of interpretation. The simpler models are all applicable to their own situations, and these situations are quite distinct from one another. In multiple capture-recapture studies, however, the methods of Fisher and Ford, Jolly and Manly and Parr may all be useful. The requirements of these methods, over and above the standard assumptions, have already been set out: Fisher-Ford assumes that survival-rate is both constant and independent of age, Jolly assumes only the latter, and Manly-Parr neither. Interpretation can be important at two stages. Firstly in deciding whether one model is undoubtedly more applicable than the others in a particular situation. And secondly, if more than one model is used on the same data set, in deciding how to reconcile differences in the answers obtained.

Manly (1970) examined the Fisher-Ford, Jolly and Manly-Parr methods, and Bishop and Sheppard (1973) examined the first two of these, in simulated situations in which survival-rate *was* constant. They found the Fisher-Ford method significantly more accurate than the others when there were low sampling intensities (less than 12%), low survival-rates (around 0.5), and/or small population sizes (less than 1000). Manly also found that the Fisher-Ford method was fairly robust unless the variations in survival-rate were quite large. It therefore seems fair to conclude that the Fisher-Ford method can be used so long as the tests for variation in survival-rate (Chapter 4) do not prove significant. It will be particularly useful whenever sampling intensity, survival-rate or population size are low.

From an opposite point of view: whenever there are sufficient data, the accuracies from Jolly's method should be comparable with those from Fisher-Ford. Moreover, the methods themselves suggest that when survival-rate varies, Jolly's method should be the more appropriate of the two. This in itself has not been examined. But Manly (1970) considered simulated populations in which survival-rate varied not only with time, but also with the age of the animal. He found that Jolly's method was considerably more robust than Fisher-Ford. In fact, except in the case of high infant mortality, Jolly's method compared favourably with Manly-Parr. This is in agreement with a point made by Seber (1973): that, although Jolly's method assumes age-independent survival, it will be largely unaffected by age-*dependent* survival, so long as survival is independent of mark status, and capture-probability independent of age. Thus Jolly's method is appropriate whenever there are sufficient data, except in cases of *strongly* age-dependent survival.

Manly and Parr (1968), themselves, noted that their method would be inaccurate unless their W_i-term exceeded 10; and Manly (1970) confirmed this, suggesting the necessity for 'moderately large' samples (apparently around 40%). Manly also showed, however, that in cases of strongly age-dependent survival, especially those of high infant mortality, the Manly-Parr method was the most accurate. These two facets characterize the situations in which the method is appropriate.

Fisher-Ford assumes more than Jolly, which assumes more than Manly-Parr. But Manly-Parr requires more extensive data than Jolly, which requires more extensive data than Fisher-Ford. If the data are sparse, and survival-rate both constant and age-independent, then Fisher-Ford is obviously the most applicable method. If the data are extensive, and survival-rate both variable and age-dependent, then Manly-Parr is appropriate. But there will be many situations in which the pros and cons are shared more evenly. It should be noted, for instance, that the more restrictive models are both fairly robust when their assumptions are violated. Thus, Jolly is preferable to Fisher-Ford only if survival-rate varies *significantly*; and Manly-Parr preferable to Jolly only if survival is *strongly* age-dependent. We must recognize, therefore, that it will often be impossible to know that one model is unequivocally best. At such times the reconciliation of answers from different models becomes important.

The above argument suggests that there should never be more than two models used for the same multiple capture-recapture data. It is easy, nevertheless, to imagine situations in which the choice between Fisher-Ford and Jolly, or between Jolly and Manly-Parr, is impossible to make. If the results from the two models are in agreement, there is no problem. But if there are differences, then these must be reconciled. Occasionally this may be straightforward. Bishop and Sheppard (1973), for instance, showed that when survival-rate is constant but data are scanty, Jolly's Method consistently over-estimated both N_i and ϕ_i, as compared to both the true values and those obtained by the Fisher-Ford method. Thus, if these two methods have been applied to the same data (because it is impossible to choose between them), and the Jolly estimates are consistently greater than those from the Fisher-Ford method, it seems reasonable to place greater reliance in the latter. Similarly, Manly (1970) showed that, in cases of high infant mortality, Jolly's method consistently overestimated both N_i and ϕ_i as compared to the Manly-Parr method.

Usually, however, there will be no simple rules for reconciliation. There is, in fact, little advice that can be given. One criterion that should *not* be used is agreement with preconceptions. It should be remembered, however, that when data are too scanty for a model, that model will tend to behave unreasonably. Imagine, for instance, a comparison between results from Jolly and Manly-Parr in which the latter show wild fluctuations in $\hat{N}_i$ and $\hat{\phi}_i$. If these fluctuations are, biologically, quite unreasonable, it seems sensible to conclude that there are insufficient data for the Manly-Parr method. The Jolly results should, therefore, be preferred.

When there is *no* basis on which to distinguish between results, and the assumptions have all been satisfactorily checked, the conclusion that our best estimate lies between the two results seems unavoidable. Ultimately, as always, interpretation is the investigator's responsibility.

4 Testing the Assumptions

Armed with his repertoire of capture-recapture models, the animal ecologist can set out to estimate the parameters of natural populations. He will have two aims in mind. Firstly, he must try to collect as much data as possible so as to maximize the statistical precision of his estimates. But he must also try to ensure that the assumptions of his model are valid. Unfortunately, these two aims – both admirable – are in opposition to one another. Data collected with a painstaking regard for a model's assumptions are likely to be comparatively sparse; data collected with the emphasis on quantity are likely to involve considerable violation of assumptions. This will be illustrated in the next chapter. For now, we may conclude that the methods adopted by the successful ecologist will be based, essentially, on compromise.

In practice, as we have seen, no results are ever absolutely precise, and the assessment of precision is often difficult. Similarly, no actual study ever satisfies all assumptions in the very finest of detail. The statistical point can be accommodated by accepting that absolute precision is rarely if ever required (Chapter 1). The practical, biological point will be the subject matter of the rest of this chapter.

It would be easy to assert that because capture-recapture assumptions are never truly valid, the models should never be used. Unfortunately this would often deny us any information at all; there is usually no reliable alternative to capture-recapture which makes *more* realistic assumptions. The answer is to view 'validity' from a more utilitarian standpoint. The greater the discrepancy between theory and practice (the grosser the violation of assumptions), the less reliable the results will be. Yet studies differ in purpose, and require results of differing reliability. The utilitarian view, then, is that assumptions are valid so long as they do not lead to results which are so unreliable as to be useless. Ultimately, therefore, the validity of assumptions will depend on the purpose of the investigation concerned. Statistical tests, which assess the validity of assumptions using the original field data, take up most of the present chapter. Many of them are also described by Seber (1973). The implication of the preceding argument is that they should be interpreted in the context of the study in question.

It may often be preferable, however, to examine assumptions by using subsidiary field experiments specifically designed for the purpose. It may be suspected, for instance, that animals with a large number of marks have an increased (or decreased) chance of being recaptured. An area distinct from the main experimental plot can be chosen, and, on a particular day, an equal number of animals given one mark, two marks, three marks, and so on. These animals can be recaptured subsequently in the usual fashion. There may then be consistent differences in the numbers recaptured from different marking-groups. This would be good evidence that an animal's chances of recapture are influenced by the number of marks it has. It is, of course, impossible to specify the nature of all such subsidiary experiments. This will depend on the ecology of the particular species, the circumstances of the particular study, and the suspicions of the particular investigator. Nevertheless, such experiments will often provide the most satisfactory means of assessing the validity of assumptions.

If assumptions *are* deemed to have been unacceptably violated, the standard conclusion will be that the study should be repeated with the method suitably modified. Sometimes, however, it may be literally impossible to repeat the study; or the degree of violation may be close to the limits of acceptability; or the study may have been undertaken to determine simply whether a population was *less* than a particular size (say 500). In any of these cases, the ability to say: 'The population appears to consist of 300 individuals, though this may well be an overestimate' will be a considerable improvement on giving no answer at all. For this reason, before the statistical tests are described, the assumptions will be re-examined and the consequences of various violations noted.

The first assumption is that all marks are permanent, and are noted on recapture. If this is violated, then there is obviously a tendency to overestimate N's, and underestimate ϕ's.

If marking increases the probability of recapture (violating the second assumption), then the marked proportion in the sample will be too high, and N's consistently underestimated. For the most part, however, the estimation of ϕ is unaffected by this violation, because only data from marked individuals are used. The exception is the Manly-Parr method, where ϕ's are underestimated because $\hat{p}$'s are used in their calculation. If marking *decreases* the probability of recapture, then the effects are just the opposite.

If marking decreases the probability of survival (assumption three), then the $\hat{\phi}$'s are good estimates for the marked sub-population, but obviously underestimate the true values for the population as a whole. (b in Jackson's positive method is similarly *over*estimated). The estimation of N, however, depends only on the sample being a true reflection of the population. $\hat{N}$'s are, therefore, unaffected by this particular violation, except in the case of a closed population (Petersen estimate and weighted mean) where N is overestimated. Once again, the converse violation has the opposite effect.

The fourth assumption is that all individuals are equally catchable. If this is violated, the $\hat{\phi}$'s will be unaffected, but the N's underestimated. The situation is different, however, if individuals with a high chance of being caught in one sample do not necessarily have a similarly high chance in another. In this case – where, in effect, there is *no correlation* between individual catchabilities on different days – neither the $\hat{N}$'s nor the $\hat{\phi}$'s are affected. Not surprisingly, if there is a correlation, the bias is greater the greater the variation in catchability.

There will often be sub-groups within the population (sexes, age-classes etc.) which differ in catchability, but within which catchabilities are constant. Once again the $\hat{\phi}$'s will be unaffected, but the N's underestimated. Moreover, if an attempt is made subsequently to sub-divide $\hat{N}$'s on the basis of the proportions of different sub-groups, then the size of the less catchable group will be underestimated still further.

Allied to the fourth assumption is the fifth: that all individuals are equally likely to survive. Generally, the violation of this is relatively unimportant; the $\hat{N}$'s will be largely unaffected, and the $\hat{\phi}$'s will reflect the average rates within the population. (Although, computing the average rate for a very variable population is itself of rather doubtful utility.) The situation will be worse if the survival differentials are associated with age. N's will be underestimated if older individuals are less likely to survive, and *vice versa*. Manly and Parr's method, in which the $\hat{N}$'s will be unaffected, is an exception to this.

Associated with this, is the assumption used in some of the models that survival rate remains constant. The consequences of violating this assumption depend, of course, on the pattern with which survival rate changes. Nevertheless, it should be noted that the $\hat{N}$'s, and not just the $\hat{\phi}$'s, will be affected by such violations.

The sixth assumption is concerned only with closed populations. If, contrary to this assumption, there is both loss and gain, the N's will be overestimated. More important than this, however, is the implicit assumption that emigration, when it occurs, is permanent. Violation of this is equivalent to the situation in which marked individuals are less liable to be captured than unmarked individuals.

The final assumption is that sampling is instantaneous, or, more realistically, that sampling periods are short in comparison with total time. As already mentioned, the direct effect of violating this assumption would be to make it impossible to associate parameter-estimates with actual moments in – or periods of – time. Further, indirect effects on the structure and random sampling of a natural population are easy to imagine.

We can now move on to the statistical tests themselves. They share a common pattern. First we decide how a particular aspect of the data should look if the assumption *had* held. We then compare this expectation with what we actually observe. Chance effects ensure that the two never coincide exactly. We must decide whether the difference between observation and expectation is significant. The tests, considered in isolation, will enable us to decide whether a difference is *statistically* significant. The onus of deciding whether a difference is *biologically* significant rests squarely on the experimenter. Statistically insignificant results can have two possible causes. Either there is truly no difference of biological significance, or there are insufficient data to uncover this difference. It is easiest to appreciate this by examining the general nature of the tests themselves.

Significance tests, as their name suggests, test the statistical significance of differences between observed and expected patterns of data. The t, F and χ^2 tests are all significance tests. In the present case, all but one of our tests are χ^2 tests, and we can concentrate on these. The precise nature of the χ^2 distribution depends on the number of *degrees of freedom*: the number of independent classes of data. If our observed and expected patterns are essentially the same, and differ only because of chance effects, our χ^2-value will be roughly the same as, or rather less than, the number of degrees of freedom. Obviously, in such a case, the probability (P) of the difference being due to chance alone is fairly high. As the difference between the two patterns increases, so does the χ^2 value. But the P-value decreases: it becomes less and less probable that the difference is due simply to chance. Examination of a table of χ^2 in, for instance, Parker (1979) will make this clear. The usual, quite arbitrary, procedure is to say that if P is less than 0.05 (a less than 5% probability of the difference being due to chance effects), then the difference is significant; and if $P < 0.01$ the difference is highly significant.

The χ^2-value, however, is not only dependent on the difference in pattern. It also reflects the amount of data involved. A comparatively small difference, obtained from a large body of data, can lead to a large (and significant) χ^2. The same (proportional) difference obtained from scanty data will lead to a much smaller (non-significant) χ^2. As already stated, a P-value in excess of 0.05 can mean either that there is no biological difference, or that the data are too scanty. An experimenter who obtains a χ^2 with $P \simeq 0.1$ from scanty data, would, therefore, be ill-advised to assume that any difference is of no biological significance – especially if he goes on to conclude that the assumptions of his model have been borne out, and that his results are wholly reliable.

Generally, the conclusion that the interpretation of these tests will often be difficult cannot be avoided. Nevertheless, they must *be* interpreted, and it may be useful to bear the following points in mind:

1) Approximate P-values should always be stated explicitly (not simply: 'P is non-significant'). The conventional significance-levels ($P = 0.05$ and $P = 0.01$) can be used; but P-values which approach significance should be taken as an indication that the validity of the assumption is in some doubt, and more data could be usefully collected.
2) No conclusion affirming the validity of an assumption should be allowed to override contrary evidence, derived either from specially-designed experiments, or from a general knowledge of a species' ecology and behaviour.
3) Tests should be interpreted in the context of the study concerned. If only approximate answers are required, then the criteria of validity can be relaxed, and *vice versa*. But *if* this is being done, one must state explicitly that it *is* being done.
4) Most important of all: the onus of interpretation rests with the experimenter. Statistical tests can never be used as a means of avoiding personal responsibility for results (Gilbert, 1973).

Needless to say, the hypocritical practice of suggesting or even proving that the assumptions are invalid, but then interpreting the results as if they held, must always be avoided (Caughley, 1977).

By way of a reminder, note that χ^2's are usually computed using the general equation:

$$\chi^2 = \sum \frac{(\text{observed} - \text{expected})^2}{\text{expected}}$$

Further reminders concerning the use of χ^2 are given in the following tests when they are first appropriate.

4.1 Constancy of survival-rate (Fisher-Ford)

The Fisher-Ford method relies on the assumption of a constant survival rate. We can use the Common Blue data again to examine two tests of this assumption.

4.1.1 Day-to-day differences

Day-to-day differences in the survival rate of the whole population can be examined by comparing individual $\hat{A}_i m_i$'s with their corresponding $\sum_j m_{ij}(i-j)$'s. Although the sums of these two quantities

Table 4.1 Observed and expected days survived (Common Blue data, Fisher-Ford method).

Day, i	$\hat{A}_i m_i$	$\sum_j m_{ij}(i-j)$
4	7.36	6
5	47.66	41
6	97.97	105
8	167.46	176
9	131.32	145
12	90.37	91
13	168.03	152
14	72.85	67

have already been equated, large inequalities on particular days will indicate differences in survival rate (Table 4.1).

Note that day *2* has been excluded, because the two terms must always be the same.

The $\sum_j m_{ij}(i-j)$'s were actually observed; the $\hat{A}_i m_i$'s are what would be expected if ϕ *had* remained constant. The two sequences can be compared using a χ^2 test with seven degrees of freedom (eight days *minus* one):

$$\chi^2 = 5.55$$
$$P \simeq 0.6$$

The assumption that ϕ remained constant was, therefore, well-justified from this point of view.

4.1.2 Age-related differences

If 0.803 of the population survive over one day, then $0.803^2 = 0.645$ should survive over two days, and so on. In other words, survival rate should be independent of age. We can test whether or not this is so. Observeds are obtained by grouping together all M_{ij}'s with $(i-j)$ equal to (say) 1, giving the sum total of marks caught one day after release:

$$m_{2\,1} + m_{5\,4} + m_{9\,8} + m_{13\,12} + m_{14\,13} = 49$$

This is then repeated for all $(i-j)$ values from *2* to *13*.

To calculate expecteds, another trellis must be constructed with an expected value corresponding to each $m_{i\,j}$. Consider, for instance, an 'expected $- m_{95}$'. 50 individuals were released on day *5*, and of these $50 \times 0.803^{(9-5)}$ should survive until day *9*. On day *9*, n_9/N_9 of the population are caught, and the expected number of day *5* marks caught on day *9* is given by:

Table 4.2 Trellis of m_{ij}–expecteds (Common Blue data, Fisher-Ford method).

Day of capture, *i*	Day of release, *j*								
	1	*2*	*4*	*5*	*6*	*8*	*9*	*12*	*13*
2	5.86								
4	1.66	2.06							
5	5.82	7.25	3.37						
6	6.68	8.32	3.87	20.08					
8	4.91	6.11	2.84	14.76	18.74				
9	2.56	3.19	1.48	7.69	9.77	15.45			
12	0.75	0.94	0.44	2.26	2.87	4.54	5.44		
13	1.09	1.36	0.63	3.28	4.17	6.60	7.90	4.58	
14	0.37	0.46	0.21	1.11	1.41	2.23	2.67	1.55	2.57

$$50 \times 0.803^4 \times \frac{50}{134.9} = 7.69$$

and generally:

$$\text{`}m_{ij}\text{–expected'} = \frac{r_j \phi^{(i-j)} n_i}{N_i}$$

The trellis in Table 4.2 can, therefore, be constructed.

Expecteds are then summed in the same way as observeds, leading to Table 4.3.

Each age can then be taken in turn and subjected to a χ^2 test with one degree of freedom. Following this, a particular property of χ^2 can be utilized. Several independent χ^2's can be added together, and the

Table 4.3 Observed and expected recaptures of various ages (Common Blue data, Fisher-Ford method).

Age (days)	Observed	Expected
1	49	51.91
2	23	26.22
3	34	38.88
4	42	37.11
5	15	17.43
6	15	11.21
7	14	14.53
8	10	7.69
9	4	1.74
10	1	1.15
11	2	2.11
12	1	1.55
13	0	0.46

significance of this single value tested by also adding their degrees of freedom. The advantage of this is that several near-significant χ^2's can be compounded to give a single significant χ^2. Caution should be exercised, however, because the different, independent χ^2's may indicate quite different effects. Adding a χ^2 from a near-significant excess to one from a near-significant deficit, for instance, may give a significant result – but this significance will be entirely spurious.

In the present case, all ages from 9 to 13 days must be grouped so that the expected exceeds 5. This leads to nine χ^2's, all of which are far from significant (*P*-values of approximately 0.7, 0.55, 0.45, 0.45, 0.6, 0.25, 0.9, 0.45 and 0.7). The lack of any apparent trend makes addition of these χ^2's inadvisable. Once again, the assumption of a single, constant survival rate has been supported.

4.2 Differences between sub-groups

It is desirable, whenever possible, to analyse results from different sub-groups separately, but this may be impossible because of insufficient data. In such cases, a test of whether the survival-rates and/or the catchabilities of the different sub-groups are significantly different should be carried out. It would be sensible, in fact, to carry out such tests on good data-sets, but then apply the results to subsequent experiments where sub-groups *cannot* be analysed separately.

4.2.1 Petersen estimate

Imagine a population containing several recognizable and mutually exclusive subgroups $A, B \ldots X$, with survival probabilities $\phi_A, \phi_B \ldots \phi_X$, and probabilities of capture $p_A, p_B \ldots p_X$. Our expectation is based on the hypothesis that individuals from different sub-groups caught in the first sample, are equally likely to reappear in the second:

$$H_o: \phi_A p_A = \phi_B p_B = \ldots \phi_X p_X$$

This expectation can be compared with what is observed by drawing up a χ^2 *contingency table*. In such a table the data are classified in two ways. Firstly, each individual captured on day *1* is assigned to one of the sub-groups; and secondly, each of these individuals is classified as either recaptured or not-recaptured on day *2*. For instance, of $r_{1\,A}$ sub-group A individuals released on day *1*, $m_{2\,A}$ are recaptured on day *2*, and $r_{1\,A} - m_{2\,A}$ are not recaptured. The following contingency table can be drawn up:

	Sub-group				
	A	*B*	....	*X*	Total
Recaptured	m_{2A}	m_{2B}	....	m_{2X}	m_2
Not recaptured	$r_{1A}-m_{2A}$	$r_{1B}-m_{2B}$	....	$r_{1X}-m_{2X}$	r_1-m_2
Released	r_{1A}	r_{1B}	....	r_{1X}	r_1

Every one of these quantities is observed – the last row (*Released*) and last column (*Total*) being the totals of the cells in the preceding rows and columns. We must derive expecteds for each of these cells.

Our original hypothesis amounts to the assertion that the chances of an individual appearing in the *Recaptured* (or *Not recaptured*) row are independent of which sub-group it belongs to. In other words, since the overall probability of being recaptured is m_2/r_1, and since the probability of an individual chosen at random from the first sample being in sub-group *A* is $r_{1\,A}/r_1$, the probability of being both in sub-group *A and* being recaptured is $\frac{r_{1\,A}m_2}{r_1^2}$. The expected *number* of sub-group *A* recaptures is, therefore, given by:

$$\frac{r_{1\,A}m_2}{r_1^2}\,.\,r_1 = \frac{r_{1\,A}m_2}{r_1}$$

This is the expected value corresponding to $m_{2\,A}$. The process can be repeated for each cell: multiplying the row total by the column total and dividing by the grand total.

We now have a matrix of observeds, and a matrix of expecteds (with the 'totals' row and column being ignored). A single χ^2 value can be computed. Note that the number of degrees of freedom is given by $(rows-1)\times(columns-1)$: once again, the number of *independent* classes of data.

The same contingency table can be used to test for differences in capture probabilities alone, by assuming a constant survival probability. It is then testing whether sampling is random with respect to the different sub-groups. If, conversely, the capture probabilities are assumed constant, the test is for constancy of survival probability.

Example. Shells of the land snail *Cepaea nemoralis* can be yellow, brown or pink:

Observed:

	Yellow	Brown	Pink	Total
Recaptured	14	8	10	32
Not recaptured	39	16	10	65
Released	53	24	20	97

Expected: ($m_{2\ yellow} = 53 \times 32/97$ etc.)

	Yellow	Brown	Pink
Recaptured	17.48	7.92	6.60
Not recaptured	35.52	16.08	13.40

$(O-E)^2/E$:

	Yellow	Brown	Pink
Recaptured	0.69	0.00	1.75
Not recaptured	0.34	0.00	0.86

$$\chi^2 = 3.64 \text{ with two degrees of freedom } (P \simeq 0.19)$$

The difference is not significant, and our hypothesis – that yellow, brown and pink individuals caught on day *1* are equally likely to reappear on day *2* – has withstood this test. A *P*-value of 0.19, obtained from only 97 individuals, does suggest, however, that it would be sensible to collect more data.

4.2.2 *Multiple capture-recapture*

A procedure can be used for multiple capture-recapture, which is entirely analogous to the one used for the Petersen estimate. Of r_j individuals released on day *j*, m_{ij} are recaptured on day *i*, and $\sum_i m_{ij}$ are recaptured on all subsequent days. (m_{ij} refers to individuals, not marks). If the r_j's and $\sum_i m_{ij}$'s are assigned to the sub-groups *A*, *B*...*X*, the following contingency table can be constructed:

	Sub-group				
	A	B	....	X	Total
Recaptured	$\sum_i m_{ijA}$	$\sum_i m_{ijB}$	....	$\sum_i m_{ijX}$	$\sum_i m_{ij}$
Not recaptured	$r_{jA}-\sum_i m_{ijA}$	$r_{jB}-\sum_i m_{ijB}$	....	$r_{jX}-\sum_i m_{ijX}$	$r_j-\sum_i m_{ij}$
Released	r_{jA}	r_{jB}	....	r_{jX}	r_j

This can be repeated for each j, and the χ^2's (and degrees of freedom) summed.

Essentially, our hypothesis is concerned, once again, with both survival and capture probabilities. As with the Petersen estimate, however, we can use it to test whether sampling is random with respect to the different sub-groups, by assuming that survival-rates are constant. The hypothesis is then that individuals in different sub-groups are equally catchable. Note that if the survival-rates are assumed not only to be constant but also all equal to 1, then the marks at risk will accumulate. The r_j's of the contingency table may then be replaced by m_i's.

Example. The data in Table 4.4 refer to males and females of the fruit-fly *Drosophila subobscura* studied in a British woodland.

For $j = 1$, observed:

	Male	Female	Total
Recaptured	6	9	15
Not recaptured	18	12	30
Released	24	21	45

$$\chi^2 = 1.61$$
$$\text{For } j = 2, \quad \chi^2 = 8.01$$
$$\text{For } j = 3, \quad \chi^2 = 2.37$$

The total χ^2, with three degrees of freedom, is 11.99.

The total χ^2, and that for $j = 2$ are highly significant ($P<0.01$); those for $j = 1$ and $j = 3$ are not. Overall we must conclude that males and females are not equally likely to be recaptured (although we cannot say whether this is a result of differential survival, differential catchability or a combination of the two).

Table 4.4 Capture-recapture data for the fruit-fly *Drosophila subobscura*.

	Male				Female			
Day, i	r_i	m_{ij}			r_i	m_{ij}		
1	24				21			
2	65	2			52	5		
3	76	2	3		67	1	5	
4		2	8	5		3	16	12

4.3 Sub-group survival (selective value)

The sub-group test for the Petersen estimate can be converted to either a catchability or a survival test by assuming that either survival or catchability is constant. It has already been noted that the same thing can be done to derive a catchability test using multiple capture-recapture data. To test for differentials in survival-rate, however, a different test can be used (Manly, 1973).

Imagine a population containing two morphs: *I* and *II*. The (constant) survival-rate of the first is ϕ_I. That of the second can be denoted by $\phi_I W$. Any difference in survival-rate can be examined by computing a confidence interval for *W*.

Of the r_{jI} morph *I* individuals released on day *j*, $r_{jI}\ \phi_I$ survive until day $(j+1)$, $r_{jI}\ \phi_I^2$ survive until day $(j+2)$, and, generally, $r_{jI}\ \phi_I^{i-j}$ survive until day *i*. On day *i*, m_{ijI} of these are recaptured. The situation for the r_{jII} morph *II* individuals is just the same: $r_{jII}(\phi_I W)^{i-j}$ individuals survive until day *i*, when m_{ijII} are recaptured.

Since the catchabilities of the two forms are the same:

$$\frac{m_{ijI}}{r_{jI}\ \phi_I^{i-j}} = \frac{m_{ijII}}{r_{jII}(\phi_I W)^{i-j}}$$

The ϕ_I^{i-j}'s cancel, and the equation rearranges to:

$$m_{ijII} - \frac{r_{jII} W^{i-j} m_{ijI}}{r_{jI}} = 0$$

In other words, for every combination of *i* and *j* this equation ought to hold; and the *W*-value we require is the one which satisfies it. It is certain, however, that each *ij*-combination will give a slightly different answer, and that our most accurate single answer will be obtained by pooling data.

We can start by pooling those figures for which $(i-j)$ takes a particular value. For instance, if $i-j = 1$, the combinations *1,2*; *2,3*; *3,4* and so on, should all satisfy the equation:

$$m_{ijII} - \frac{r_{jII} W m_{ijI}}{r_{jI}} = 0$$

So, therefore, should the equation:

$$\sum_{j} \left\{ m_{ijII} - \frac{r_{jII} W m_{ijI}}{r_{jI}} \right\} = 0$$

Note that in this equation j can take any value, but i is always subject to the condition that $i-j=1$.

We are left now with a series of such equations: one for $i-j=1$, one for $i-j=2$, and so on. These, too, must be pooled. But we are considering a parameter which accumulates its effects with time. Estimates gained from longer periods (i.e. when $i-j$ is large) are likely, therefore, to be relatively more accurate. For this reason we weight each equation by its $(i-j)$-value in our final sum:

$$f(\hat{W}) = \sum_{i-j} \left\{ (i-j) \sum_{j} \left(m_{ijII} - \frac{r_{jII} \hat{W}^{i-j} m_{ijI}}{r_{jI}} \right) \right\} = 0$$

Our $\hat{W}$-value will satisfy this equation, and we must find it by trial and error – just as we found ϕ in Jackson's negative method and the Fisher-Ford method.

The standard error for $\hat{W}$ is given by:

$$\mathrm{SE}_{\hat{W}} = \frac{\hat{W}}{g(\hat{W})} \left\{ \frac{g(\hat{W})}{r_{jII}} + \frac{\sum_{i-j} R_{ij}(i-j)\hat{W}^{i-j}}{r_{jI}} - \left(\frac{1}{r_{jII}} + \frac{1}{r_{jI}} \right) \left(\sum_{i-j} R_{ij} \right)^2 \right\}^{\frac{1}{2}}$$

$$\text{where } g(\hat{W}) = \sum_{i-j} (i-j) R_{ij}$$

$$\text{and } R_{ij} = \frac{(i-j)\hat{W}^{i-j} \left(\sum_{j} m_{ijI} \right)}{r_{jI}}$$

We can then take the confidence interval of $\hat{W}$ as:

$$(\hat{W} + 2\mathrm{SE}_{\hat{W}}) - (\hat{W} - 2\mathrm{SE}_{\hat{W}})$$

If the survival-rates of morphs I and II are the same ($\phi_I = \phi_I W$), then W should equal unity. Thus, if the confidence interval of $\hat{W}$ does not include unity, the survival-rates of the two morphs can be taken as different.

This in itself represents a test for differential survivorship between sub-groups of a population studied by multiple capture-recapture. But the method was developed for a different purpose. Manly

Table 4.5 *Amathes glareosa* recapture data (Kettlewell *et al.* 1969).

i	*typica* (I)		*edda* (II)	
	r_{iI}	m_{i1I}	r_{iII}	m_{i1II}
1	2050		2235	
$2\frac{1}{4}$		20		34
$3\frac{1}{4}$		16		26
$4\frac{1}{4}$		11		19
$5\frac{1}{4}$		6		11
$6\frac{1}{4}$		3		6
$7\frac{1}{4}$		2		3
$8\frac{1}{4}$		1		2
$9\frac{1}{4}$		1		2
$10\frac{1}{4}$		0		1
$11\frac{1}{4}$		1		1
$12\frac{1}{4}$		0		1
$13\frac{1}{4}$		0		0
$14\frac{1}{4}$		0		1

considered the case where the morphs are distinguished not by sexual or age-specific differences, but rather by the permanent, genetically-determined differences exemplified by melanic and non-melanic moths of a single species. He then assumed that differential survivorship inferred differential mortality. The morph with the lower mortality-rate will, therefore, live longer and, on average, contribute more to future generations. It will, from an evolutionary point of view, be at a selective advantage. In such a case the selective value of morph *I* can be taken as unity, and W becomes the selective value of morph *II*. The method then tests whether morph *I* is at a selective advantage ($\hat{W}$ significantly less than 1); or morph *II* is at a selective advantage ($\hat{W}$ significantly greater than 1); or the selective values of the two morphs are not significantly different from one another (the confidence interval of $\hat{W}$ includes 1). Not surprisingly, the method is of considerable importance to the ecological geneticist.

Example. The data in Table 4.5 are those reported by Kettlewell *et al.* (1969) for the moth *Amathes glareosa*, which exists as two distinct morphs: *typica* and *edda*. This is a simplified example in that marked individuals were released on only one occasion: $j = 1$ only. It is slightly complicated, however, by the fact that individuals were recaptured somewhat later in the day than the release. Recaptures are, therefore, taken as occurring after $2\frac{1}{4}$ days, $3\frac{1}{4}$ days, and so on.

We begin by assuming $\hat{W} = 1$, and construct Table 4.6. Note, once again, that this table is much simplified, because j only takes the value

Table 4.6 Preliminary computations for Manly's (1973) method–see text. Data from Kettlewell *et al.* (1969).

$i-1$	$\frac{r_{1il}\hat{W}^{i-1}m_{i1}}{r_{1l}}$	$(i-1)(m_{i1ll}-\frac{r_{1il}\hat{W}^{i-1}m_{i1l}}{r_{1l}})$
1.25	21.9	15.13
2.25	17.5	19.13
3.25	12.1	22.43
4.25	6.6	18.7
5.25	3.4	13.65
6.25	2.3	4.38
7.25	1.2	5.8
8.25	1.2	6.6
9.25	0	9.25
10.25	1.2	−2.05
11.25	0	11.25
12.25	0	0
13.25	0	13.25
		137.52 = $f(\hat{W})$

1. In fuller cases, each of the terms in column three would themselves be sums, taken over the range of j-values.

In this case $f(\hat{W})$ greatly exceeds zero, and the true value of $\hat{W}$ must be greater than 1. We therefore repeat the procedure for a $\hat{W}$-value of 1.2, giving a value for $f(\hat{W})$ of −138.95. It is sensible at this point to plot our two points on a graph (Fig. 4.1), which suggests that a $\hat{W}$-value of 1.11 should be appropriate. If this is tried, a $f(\hat{W})$ of 18.45 is obtained. The true $\hat{W}$ is, therefore, slightly greater than 1.11. This suggests a $\hat{W}$ of 1.12, which leads to a $f(\hat{W})$ of 4.35.

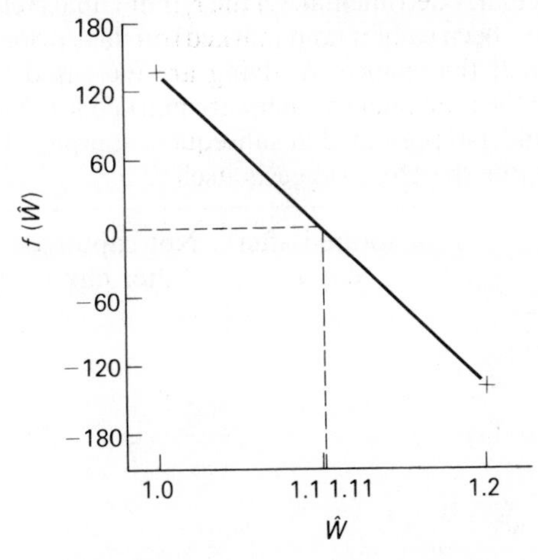

Fig. 4.1 Preliminary estimation of $\hat{W}$ for Manly's (1973) method.

We can only get closer to our true $\hat{W}$ by dealing with more than two decimal places. It is preferable to accept 1.12 as the selective value of the *edda* morph as compared to the *typica*.

The standard error of this value if 0.05, and the confidence interval for $\hat{W}$ is 1.02–1.22.

This result can be interpreted as meaning either (a) that the survival-rate of the *edda* morph is significantly greater than that of the *typica*; or (b) that the relative selective values of the two morphs of *Amathes glareosa* are *typica*:1, *edda*:1.12, and that this difference is significant.

4.4 Effects of marking

It is sensible to test the assumption that capturing and marking have no effect. This assumption can be split into two: that marked and unmarked (caught and uncaught) individuals can be treated as equivalent; and that all marked individuals, however many times they have been caught, can themselves be treated as equivalent. Each of the following tests utilizes aspects of the original capture-recapture data, and tests only the second of these assumptions. The essential reason for this is that there *is* no useful method for testing the first assumption. It seems reasonable to suggest, however, that if the second assumption holds, the first will too.

4.4.1 Initial marking mortality

In certain experiments, it may be suspected that the experience of being marked for the first time – or, in the case of individual marks, being caught for the first time and marked for the one and only time – is particularly detrimental. Of the r_i individuals released on day i, some will have been caught (and marked) on days prior to day i, and some will not. If the chances of dying are increased by the initial marking, then the individuals which were marked for the first time on day i will be under-represented in subsequent samples. The following contingency table therefore suggests itself:

	Captured after day i	Not captured after day i	Total
First captured on day i and released			$r_i - \sum_j m_{ij}$
Recaptured on day i and released			$\sum_j m_{ij}$
Captured on day i and released	y_i	$r_i - y_i$	r_i

χ^2's (and their degrees of freedom) can then be summed for different i's.

Example. The following data were obtained by Manly and Parr (1968) for the six-spot Burnet moth (*Zygaena filipendula*):

Day *1*: 57 captured, marked (*green*) and released.
Day *2*: 52 captured (25 *g*), marked (*white*) and released.
Day *3*: 52 captured (8 *g*, 9 *w*, 11 *gw*), marked (*blue*) and released.
Day *4*: 31 captured (2 *g*, 3 *w*, 4 *b*, 5 *gb*, 1 *wb*, 2 *gwb*), marked (*orange*) and released.
Day *5*: 54 captured (1 *g*, 2 *w*, 7 *b*, 5 *o*, 4 *gw*, 2 *gb*, 2 *go*, 4 *wb*, 1 *wo*, 1 *bo*, 5 *gbo*, 1 *gwbo*).

For day *2*, of 52 released, 25 were marked and 27 unmarked. Of these, 15 of the marked moths were subsequently recaptured (11 on day *3* and 4 on day *5*), and 14 of the unmarked moths (9 on day *3*, 3 on day *4* and 2 on day *5*). The following table of observeds can be drawn up:

	Captured after day *2*	Not captured after day *2*	Total
First captured on day *2* and released	14	13	27
Recaptured on day *2* and released	15	10	25
Captured on day *2* and released	29	23	52

leading to a χ^2 of 0.34 ($P \simeq 0.6$).

A similar contingency table for day *3* leads to a χ^2 of 0.09 ($P \simeq 0.75$); while, for day *4*, χ^2 is 1.63 ($P \simeq 0.2$). The total χ^2 (with three degrees of freedom) is 2.06 ($P \simeq 0.6$).

All of these *P*-values suggest that being caught and marked for the first time did not affect the moths significantly.

4.4.2 Independence of mark status

Perhaps the most predictable effect of being caught and marked, is that individuals which have been caught (and possibly marked) many times, are either more or less likely to be subsequently recaptured than individuals caught less often. Of the r_i individuals released on day i, y_i will be caught subsequently. Of these r_i individuals, $r_{0\,i}$ will have never been caught prior to day i, r_{1i} will have been caught once prior to day i, and, generally, r_{xi} will have been caught x times previously. Of these, y_{xi} will be subsequently recaptured. The following contingency table can be drawn up:

	Times caught previously						
	0	*1*	...	x	...	$i-1$	Total
Recaptured	y_{0i}	y_{1i}	...	y_{xi}	...	$y_{i,i-1}$	y_i
Not recaptured	$r_{0i}-y_{0i}$	$r_{1i}-y_{1i}$	...	$r_{xi}-y_{xi}$	...	$r_{i,i-1} - y_{i,i-1}$	$r_i - y_i$
Released	r_{0i}	r_{1i}	...	r_{xi}	...	$r_{i,i-1}$	r_i

χ^2's (and degrees of freedom) can then be added for different values of i. As with the sub-group tests, our hypothesis here is concerned with both catchability and survival. Each can be tested alone by assuming that the other remains constant.

A particular difficulty should be mentioned. An individual with several marks could quite plausibly have a reduced probability of survival, but also an increased chance of being captured. The nature of our hypothesis:

$$H_o: \phi_{0i}P_{0i} = \phi_{1i}p_{1i} \dots = \phi_{xi}p_{xi}$$

– means that these two effects will tend to mask one another. The hypothesis will then be accepted, even though both N_i and ϕ_i are underestimated. Fortunately, the test which follows this is concerned only with catchability.

Example. Again, Manly and Parr's data on *Z. filipendula* can be used. For instance, of the day *3* sample 24 individuals had never been caught before, and of these 11 (4+7) were subsequently recaptured; 17 (8+9) individuals had been caught once of which 12 (5+1+2+4) were subsequently recaptured; and 11 had been caught twice of which 2 were subsequently recaptured. The following table of observeds can be drawn up:

	Times caught previously			
	0	*1*	*2*	Total
Recaptured	11	12	2	25
Not recaptured	13	5	9	27
Released	24	17	11	52

leading to a χ^2 (with two degrees of freedom) of 7.46 ($P \simeq 0.025$). Similarly, for day *2*, a χ^2 (with one degree of freedom) of 0.34 ($P \simeq 0.6$) is obtained. For day *4*, the columns for $x = 1, 2$ and 3 must be pooled, giving a χ^2 (one degree of freedom) of 1.63 ($P \simeq 0.2$). The total χ^2 (four degrees of freedom) is 9.43 ($P \simeq 0.05$).

These results are difficult to interpret. The χ^2 for day *3*, and the total χ^2 are both significant; but the reasons for this are rather obscure. On day *3*, for instance, there was a tendency for individuals with one previous mark to reappear more often than expected, but for those with two previous marks to reappear less often. On other days there was a general tendency for previously-marked individuals to reappear more often than expected. A further complication, of course, is that this test compounds two biologically-distinct factors: survival and catchability.

Overall, we can do no more than admit that serious doubt has been thrown on the assumption that individuals are unaffected by being marked. Our only conclusion is that we need more information on *Z. filipendula* before we can assess the reliability of the capture-recapture estimates.

4.5 Random sampling

As mentioned above, this test – developed by Leslie (1958) – is concerned only with the catchabilities of marked individuals. Our hypothesis is that the marked population as a whole is sampled at random. It should be noted, however, as pointed out by Roff (1973b), that this test is effectively incapable of distinguishing between random sampling, and random sampling within sub-groups which themselves differ in catchability.

Consider a group of G individuals known to be alive and susceptible to capture between days j and $j+t$. In other words, these are individuals caught both on or before day j, and on or after day $j+t$. (We must assume, as usual, that there is no temporary emigration.) Of these individuals, g_i are caught on day i – where i lies between j and $j+t$. The mean number of captures per individual is given by:

$$\mu = \frac{\sum_i g_i}{G}$$

and the variance of this by:

$$\sigma^2 = \mu - \frac{\sum_i g_i^2}{G^2}$$

This, at least, is what we *expect* on the assumption of random sampling. But if, between days j and $j+t$, f_x individuals from G are caught x times, then the *observed* variance is:

$$\frac{\sum_x f_x(x-\mu)^2}{G-1}$$

If sampling *is* random, then the observed and expected variances should be equal, and dividing one by the other should give 1. Similarly:

$$T = \frac{\sum_x f_x(x-\mu)^2}{\sigma^2}$$

should be equal to $(G-1)$. Leslie showed that T approximated to χ^2 with $(G-1)$ degrees of freedom. Thus, as usual, a significant χ^2 signifies a significant departure from our hypothesis. Following Leslie, we can consider the approximation satisfactory when $G>20$, and there are at least three samples between days j and $j+t$.

Example. Once again, we can use Manly and Parr's *Z. filipendula* data.

The only satisfactory interval is between days *1* and *5*, in that there were three intermediate samples: on days *2*, *3* and *4*. During this time there were 15 individuals *known* to be alive; that is, caught on day *1* and given a green mark, and recaptured on day *5 with* a green mark. Consequently, G (= 15) is smaller than is really desirable, and the test should be applied with caution. The following table can be drawn up:

i	g_i	g_i^2	x	f_x	xf_x	x^2f_x
2	5	25	0	1	0	0
3	8	64	1	8	8	8
4	8	64	2	5	10	20
	21	153	3	1	3	9
				15	21	37

$g_3 = 8$, for instance, because 8 individuals were caught on day *5* with both green and blue marks ($2gb+5gbo+1gwbo$). Similarly, $f_2 = 5$ because 5 individuals were caught on day *5* which had been marked twice since day *1* ($5gbo$).

$$\mu = 1.4$$
$$\sigma^2 = 1.4-0.68 = 0.72$$

and $$T = 10.6$$

Examination of the χ^2 table for 14 degrees of freedom indicates that the observed and expected variances are close ($P \simeq 0.7$). Consequently, even though G was small, the test does lend support to the hypothesis that sampling of the marked population was random.

Finally, it is worth repeating that the tests in this chapter are not refinements for the fastidious. They should, on the contrary, be an essential part of any attempt to apply theoretical capture-recapture models to actual capture-recapture data.

5 Practicalities

The practical aspects of capture-recapture could be described in many ways. It would be possible, for instance, to provide a catalogue of the techniques which have been employed and the species to which they have been applied. If this were done uncritically, without comment, then the catalogue, although long, would be nothing more than a list of possibilities. If each example were critically assessed, the length would be quite unmanageable. Alternatively, specific techniques could be ignored, and general points made regarding the problems and pitfalls of actual studies. But the relevance of such advice to a particular experiment would not always be apparent. Another alternative would be to concentrate on exemplary case studies, with a view to recreating the atmosphere of actual investigations. Whatever approach is taken, however, it is essential that potential investigators are aware (a) of the available techniques which are likely to prove successful, and (b) of the ways in which these may have to be modified in particular instances. The first part of this chapter is, therefore, an outline of the techniques in common use, punctuated with cautionary remarks concerning the violation of assumptions. This is followed by three case-studies (an insect, a fish and a mammal) on which the importance of context is stressed.

The single most important point to be made in this chapter is that every study is unique. Only statements of the most general nature can be applied to all capture-recapture work. It follows, inevitably, that a study cannot be designed without intimate knowledge of the species (and probably the population) involved: the greater the knowledge, the better the design, and the more accurate the results. It follows, too, that all designs should be considered provisional, in that they provide further knowledge (when the data are produced and analysed), and this can lead to design improvements. Capture-recapture results can only be reliable, therefore, if there is a body of ecological knowledge on which design can be based. But such results are also only worth having if there is an ecological background against which to interpret them. Capture-recapture studies should answer specific questions. Questions will only be worthwhile if they are asked by informed biologists.

5.1 Capture

Methods of capture have been extensively reviewed for insects (Southwood, 1966), freshwater fish (Lagler, 1971), amphibians and reptiles (Woodbury, 1956), mammals (Twigg, 1975a), and birds and mammals (Taber and Cowan, 1969). These will be outlined below. Whatever the method, however, there is one underlying assumption which is particularly pertinent: that all individuals have *an equal chance of being caught*. It is also essential, of course, that there should be *sufficient data*, and all methods of capture must be assessed in the light of these two requirements. But such assessment is only feasible if the methods are considered in the context of the study in question. The interests of statistical precision are best served by maximizing the number and efforts of the investigators: those of biological reliability by minimizing the disturbance to the population. Once again two interests conflict. The nature of the eventual compromise is the responsibility of the investigator.

Methods of capturing insects can be divided into two groups: those in which capture depends on the activity of the investigator, and those in which it depends on the activity of the insects themselves. Simplest among the methods of the first group is visual observation, where the investigator usually walks through the site, following an appropriate path, and collects all the animals he sees. A path is appropriate if it gives all individuals an equal chance of being seen, and the whole of a site should, therefore, be searched with equal efficiency. A net, glass-tube or aspirator is often used to facilitate capture once an insect is sighted, and it should be stressed that all animals should be pursued with equal determination. It is only too easy to bias samples by preferentially capturing individuals which are unusual or relatively immobile.

A minor modification of this procedure involves walking the path and collecting insects with strokes of a net, either in the air or through the vegetation. This will be useful when insects are too small or mobile to be singled out for capture. The nature of the net will, of course, depend on the medium through which it is swept.

Also included in the first group of methods are those that are more commonly used as density-estimators in their own right. The site is divided into several equal sub-areas; a proportion of these are chosen at random; and all animals within each sub-area are collected. This procedure will itself estimate density if the animals are immobile, or are in some way constrained from leaving their sub-area. But such methods are also useful in capture-recapture if, during any one sampling occasion, there are some members of the population which are uncatchable. Land-snails, for example, may be buried during one sampling occasion, but susceptible to capture during the next.

Catching all the *visible* snails in a sub-area will then leave an unknown proportion of that sub-population not captured, and the standard analyses will apply.

Methods depending on the activity of the insects themselves can be broadly divided into two further groups: those depending on interception, and those depending on attraction. A few methods combine the two. In the first group, suspended nets have been used to catch both terrestrial and aquatic insects. But from the point of view of capture-recapture, the most important interception traps are pitfall traps. These consist simply of a glass, plastic or metal jar sunk into the soil so that the mouth is level with the soil surface. Flightless animals that walk on the soil surface may then fall in and be unable to climb out. Despite their shortcomings as a means of sampling a habitat, pitfall traps may still be useful for collecting a variety of animals for capture-recapture, especially spiders and carabid beetles.

Among attraction-traps, the most commonly used is the light trap; even in 1966 Southwood was able to refer to 'several hundred' relevant references. Traps vary in the wavelength of the light which they emit – visible, ultra-violet, or a combination of the two – but all seem to operate by interfering with the insects' normal photic orientation. They seem to be attracted from afar, but repelled by the high light intensity immediately adjacent to the lamp. Consequently most traps have an arrangement of baffles and funnels to collect the insects once they have been attracted. Macropterous and micropterous moths, Trichoptera, mirids, corixids and many Diptera have all been successfully captured with light traps.

Many other types of attraction trap have been used to collect insects, and most use as bait either a natural attractant, or some substance which mimics a natural attractant. Vertebrate 'hosts', carrion, dung, rotting and fermenting fruit, and virgin females have all been used successfully.

Whatever the type of trap, and whatever the type of animal being captured, there are several points which must be considered. Most traps have specific problems of their own: in pitfall traps, carabids tend to devour one another; virgin females only attract males, and so on. But there are many problems common to all traps. The first concerns the assumption that sampling time is relatively short. With pitfall traps and light traps, for instance, where the majority of species caught are nocturnal, it is tempting to extend the sampling time to cover the whole activity period. Although it increases the amount of data, this procedure not only violates an important underlying assumption, but also makes it difficult for the captured animals to remix with the population.

A second consideration is that a captured animal's subsequent chances of being caught should be unaffected. The twin aspects of this will be discussed when mammals and 'release' are considered.

The major consideration, however, is that sampling should be random. To some extent this will be dependent on the inherent response of the animal to the trap, irrespective of any precautions taken; and the importance of this can be assessed only by knowing the ecology of the species involved. But the design and arrangement of the traps themselves may also affect the randomness of samples. Traps, whether interceptive or attractive, will neglect individuals outside their area of influence. Such neglected individuals may, nevertheless, be within the population – especially when the traps are widely spaced. If the animals are freely mobile and the environment homogeneous, it is probable that different individuals will be involved each day, and the arrangement of traps can remain fixed. This arrangement ought, then, to ensure that each trap has, as nearly as possible, an equal area of influence. If, on the other hand, there is a suspicion that the same individuals are being neglected each time (either because of the way they are distributed, or the way they disperse), then the traps themselves should be repositioned at random each day. (Implicit in this is the assumption that there is no trap exclusion: that the capture of one individual does not reduce the chances of others being captured. With insects, this assumption is reasonable.)

The spacing of the traps is also important. If they are close together, there will be a high sampling intensity, and therefore high statistical precision. But if the number of traps is limited, this precision will be gained by reducing the total size of the area covered – possibly to an area which is so small as to be atypical. The other extreme is a large area in which there is a low sampling intensity and low statistical precision. Once again there is a necessity for compromise.

Sampling intensity will also be increased by increasing the size of each trap's area of influence: literally the size of an interception trap, or the attractiveness of an attraction trap. In the former case there are few attendant biological problems; but, highly attractive attraction traps, while undoubtedly increasing sampling intensity, might attract individuals from far beyond the experimental area. They might also leave behind attractive foci which interfere with normal behaviour between samples. Both of these are to be avoided; and so, therefore, are excessively attractive traps. These comments are, of course, equally applicable in a general sense to all other groups of animals.

Catching freshwater fish satisfactorily for capture-recapture studies is often both a difficult and fairly specialized process (Lagler, 1971). Seine netting, trapping and electric fishing may all be used successfully, but each has its own problems. Seining is only satisfactory if the fish are robust, and must be carried out skilfully; fish traps have to be lifted at frequent intervals; and electric fishing is limited to relatively shallow water, and may occasionally cause delayed mortality. The bias that occurs in all of these methods of capture has led many workers to suggest that different methods should be used on different occasions. The rationale behind this is that different biases will be uncorrelated, and the standard analyses will apply.

The catching of birds and mammals is usually a specialized process, requiring traps designed with a particular species in mind. Thus with birds a wide variety of methods have been used successfully: baited box traps, with either an automatic or a manual trigger; baited net traps in which the net is thrown over the attracted bird, often propelled by a 'cannon'; 'drive' traps in which birds are driven to the narrow end of a large funnel; mist nets – essentially interception traps; and traps in which birds are taken individually at their nest.

With mammals, too, automatic or manually-operated snares, and drive traps leading to corrals have both been used. But, especially with small mammals, it is the baited box trap which has proved most successful. Delany (1974) describes three such traps, and considers the 'Longworth' the most popular in this country. This is an aluminium trap made of two sections: a tunnel and a nest box. The latter contains both nesting material and food, and therefore not only attracts the small mammal, but also ensures that it is warm, dry and fed after capture. (In fact, to avoid waterlogging, the nest-box should never be below the tunnel on a slope.) The attracted animal enters the tunnel, but, by stepping on a sensitive treadle at the nest-box end, closes and locks the tunnel entrance behind it. The tunnel itself is basically of a size appropriate for catching mice and voles, but can be converted for use with shrews. A variety of baits have been used: some for specific types of mammal, and some all-purpose. For mice and voles, whole grain is generally successful; and for shrews, blow-fly pupae. Other, similar traps and baits have been used with other mammals.

The comparatively sophisticated behaviour of small mammals has highlighted several problems related to capture. The first is trap-exclusion, in which the springing of a trap by one individual immediately reduces the chances of capture of other individuals in the same area. This means that if traps are distributed evenly but the animals are not, areas of high density will be sampled less intensely than areas of low density, and sampling will not be random. It also means that, after some traps have been sprung, the (territorial) animals in areas with sprung traps are less catchable than those in areas with unsprung traps. The first aspect can be countered only by an appropriate positioning of traps, which infers a detailed knowledge of the species' ecology. The second can be countered by increasing the density of traps within an area; and, more specifically, by setting two traps at each trapping-point.

The second problem is trap-shyness: a common response of small mammals (and other animals) to something new and strange in their environment. The real problem, of course, is that some animals are more trap-shy than others. It can be alleviated by allowing the animals to become accustomed to the traps' presence before the study begins. This is done most readily by leaving *unset* traps in position, with food in, for several days before trapping for the first time.

The third problem, also common, is trap addiction. Here the animals come to learn that the trap is a source of food and warmth, spoilt only by a minor trauma in the hands of the investigator. They therefore return repeatedly to the same trap. Trap addiction can be prevented to some extent by changing the position of traps each night, although this in itself may cause problems if the original trap-positions had been chosen purposefully.

Finally, the fourth problem is just the opposite of trap addiction: trap avoidance. Here the animal finds its initial experiences of trapping sufficiently traumatic to reduce its subsequent chances of recapture. The best counter to this is an improvement in sampling technique.

Whatever method of capture is used, a decision must be made concerning the location of the experimental area. Two major, related considerations are: (a) that the area should be typical of that occupied by the population as a whole; and (b) that it should not be so heterogeneous that sampling within it is non-random. White (1975) has established that these are important considerations which are all too often ignored. If possible, the area should also be located so that migration to and from it is minimized. This will increase the rate of recaptures, and also minimize temporary emigration – which all models specifically preclude. The easiest way to achieve this aim is to ensure that some, at least, of the area's boundaries are natural barriers for the species concerned.

5.2 Handling

The careful handling of animals is essential if they are to be returned to the population unharmed and largely unaffected. There is no substitute for experience, but some general comments can be made. As a general rule it is better to anaesthetize an animal than to run the risk of damaging it – and better still, if possible, to avoid both harm and anaesthetic. With insects, anaesthesia may be essential in preventing damage, especially to wings, or may be required simply because their small size makes them otherwise unmanageable. A variety of substances have been used, including chloroform, ether nitrous oxide, nitrogen and carbon dioxide. For most of these, care must be taken to avoid overdoses and to allow for after effects – but neither are likely with carbon dioxide. Its use is, therefore, to be recommended. It can be readily dispensed in the field with a 'Sparklets' cork-remover.

For obvious reasons fish present their own difficulties. It is frequently desirable, for instance, to delay their removal from water until essential by retaining them in a keep-net or pen. It is then often advisable to avoid damage in handling by dipping them in a bucket of the fish anaesthetic, MS–222.

Anaesthetics for mammals have been reviewed by Twigg (1975a), and for mammals, birds and reptiles by Taber and Cowan (1969); but for handling small mammals and birds, probably the most useful tool is a bag, into which the trap can be emptied. The animal can be held firmly in this, and then carefully removed. Birds are generally tranquilized by darkness, and a thick canvas bag is preferable. With small mammals a transparent bag is more appropriate. In either case, accomplished usage should obviate the need for an anaesthetic.

5.3 Marking

There are extensive reviews of marking methods for insects (Southwood, 1966), fish (Stott, 1971), amphibians and reptiles (Woodbury, 1956), mammals (Twigg, 1975b) and mammals and birds (Taber and Cowan, 1969). Most of these methods can be outlined under five headings: paints, dusts, mutilation, tags and bands, and radiosotopes.

5.3.1 Paints

The most successful paints for insects, and arthropods generally, are artist's oil paints and nitrocellulose lacquers (model aircraft 'dope'). The latter is usually the more readily available. They can be used, too, on snail shells, though it may be necessary to scrape off a small area of periostracum. In all cases a small spot of paint can be readily applied with any finely-pointed instrument: an entomological pin, a cocktail stick or even a grass stem. Specialist paints which reflect or fluoresce in visible or ultra-violet light may be useful with nocturnal insects. With the Lepidoptera, however, where a mark on the wing is usually required, felt-tip pens are to be recommended.

Capture-recapture methods require only date-specific marks, but there are two possible reasons for using individual-specific marks. The first is that the permanence of marks and correctness of identifications can be checked when individually-marked animals are recaptured. The second is that the period of study can be extended, whenever there are comparatively few colours and comparatively few individuals. The following marking scheme, suggested by Dobson (1962) and utilising paint spots, will make both of these points clear. All individuals are given three marks, which can be arranged in a variety of patterns:

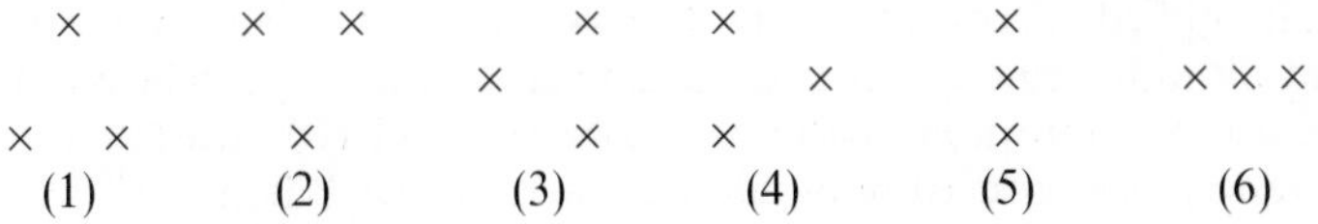

```
   ×       ×   ×          ×     ×            ×
                      ×            ×         ×         × × ×
 ×   ×       ×            ×     ×            ×
  (1)       (2)        (3)      (4)         (5)         (6)
```

and so on. If N is the number of patterns, and c the number of colours, there are Nc^3 possible combinations. Thus, with five colours and the six patterns shown, 750 individuals could be given their own mark. These can be recorded as $\mathrm{W}\overset{\mathrm{R}}{\ }\mathrm{B}$, $\underset{\mathrm{R}}{\overset{\mathrm{R}}{\ }}\mathrm{B}$, and so on. If, on average, there were only 50 new individuals caught each day, date-specific marks (with these five colours) could lead, at most, to a six-day study; whereas with Dobson's scheme sixteen days would be possible. Also, the appearance of individuals with one or two marks would be *proof* that marks were being lost. These advantages must be set, however, against a necessary increase in time and effort expended. Moreover, it should be remembered that even if individual-specific marks are used, the models require that each animal's date-specific sequence should be imagined.

As previously described, the underlying assumptions of each method can and should be tested. Nevertheless, when insects are being marked, there are several points which, if borne in mind, may prevent these assumptions being violated. The first is that marks must be permanent; which can often only be established by evidence from a subsidiary experiment. The second is that a marked animal's behaviour must be unaffected. This can most readily be ensured by a combination of biological insight and common sense. Thus, for instance, joints and sense-organs must be avoided, and amounts of paint must be small. Related to this is the third point: that a marked animal's chances of survival must be unaffected; and the fourth: that marked and unmarked individuals must be equally catchable. Both of these points depend on the behaviour of marked animals, but also on the reaction of other individuals towards them. Predators and investigators must be neither exceptionally attracted to marked individuals, nor especially likely to ignore them. Thus, marks should be applied to a surface which predators and investigators do not normally see. All of these points are, of course, applicable to other groups of animals and other methods of marking.

In studies of fish, paints, as such, are not employed, but analogous procedures such as injecting coloured dyes and liquid latex subcutaneously have been used successfully. Date-specific marks can be given by changing the position or colour of the injection.

'Painting' of birds and mammals has been rather rare. This is, perhaps, unfortunate in the case of many mammals, where the alternative is very often mutilation (see below). Appropriate dyes have been applied successfully to both dark and light pelaged mammals, and, although colours are limited, date-specific marks can be given by the use of appropriate patterns. Morejohn and Howard (1956), for instance, used a mixture of human oil-based black hair dye and 3% hydrogen peroxide in equal parts on light pelage; and a mixture of one part ammonium hydroxide (4%) to two parts hydrogen peroxide (3%) on dark. In both cases granulated soap was added to the mixture until the liquid was thick.

It is in insect-studies, however, that painting is most popular. Here, where size allows, the advantages of ease, cheapness and variety make it the most frequently used method of marking. Nevertheless, the specific position and extent of these marks should always be considered carefully.

5.3.2 Dusts

Insects which are hairy, and so small that they cannot be marked with paint, can be readily marked with very fine dusts. The most common are those which fluoresce in ultra-violet light: (zinc and cadmium sulphide). Various colours are available, and several different colours can be distinguished on the same small insect. Application can be achieved either by subjecting the insects to a cloud of dust produced by a powder dispenser in a cage; or by rolling the anaesthetized insects in a dry glass tube finely-coated with the dust. In either case it may be advisable to allow the insects to clean the dust from parts of their body before they are released. Portable ultra-violet lamps allow the whole process to be carried out in the field. The advantage of this method, is, of course, that it can be used when no other methods are possible.

5.3.3 Mutilation

This has been used only rarely as a method of marking invertebrates, usually because of size, and the dangers of affecting subsequent behaviour.

In fish, on the other hand, mutilation – in the form of fin-clipping, most commonly of the pelvic fin – is a popular method of marking. By altering the angle and position of the cut, several group-specific marks can be given. In the longer term, however, clipped fins may regenerate, and recaptures must be checked by personnel trained to recognize the altered patterns of growth. Another successful method of marking fish, which involves a small degree of mutilation, is branding with

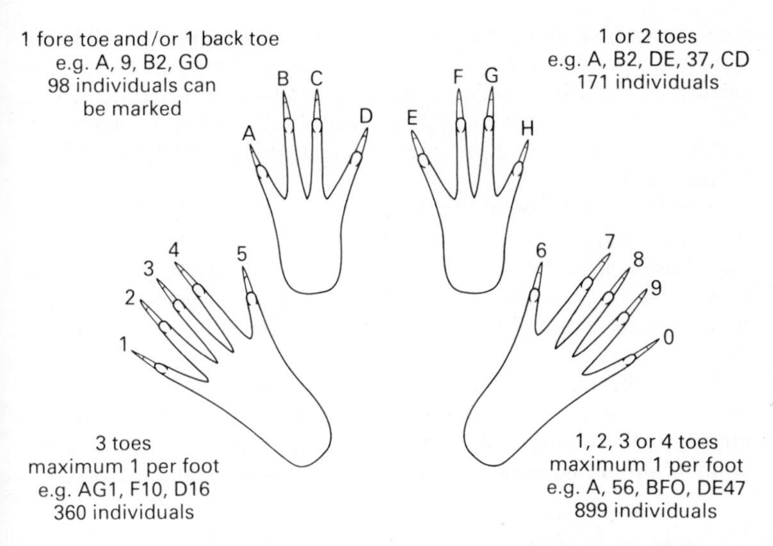

Fig. 5.1 Toe-clipping schemes for individually marking small mammals.

either hot or cold 'irons'. 'Freeze-branding' is particularly successful with salmonids.

Mutilation is also a popular method of marking mammals. Toe-clipping, tail-docking, ear-cropping, hole-punching, fur-clipping and branding have all been used. As with mutilation generally, the possible disadvantages of affecting behaviour by maiming have to be set against the ease with which the marks are applied in terms of time, effort and equipment.

Probably the commonest mutilation-mark, especially with small mammals, is toe-clipping; in which the selected toe is anaesthetized with a drop of ethyl chloride, and after a few seconds, the nail and first joint are swiftly removed with scissors. Despite its apparent cruelty, this method does appear to have relatively little effect on behaviour. (Here, and generally, it is the animal's – and not the investigator's – reaction which is important.) Toe-clipping can be used to give date-specific marks, but it is normal practice to mark mammals individually. The maximum number marked depends on how many toes are clipped per foot. Some examples are given in Fig. 5.1.

5.3.4 Tags and bands

This is another method of limited use with invertebrates, because of their size. Fish, on the other hand, have often been individually marked with tags or labels. These are usually metal or plastic, connected to the body by a wire loop. Their optimal position tends to vary, depending on the size of the fish and the strengths of the different parts of its body, but, generally, firmly-attached labels are better than those that trail. In either case, however, the method is expensive in time, effort and equipment, and may well affect subsequent behaviour. When group-specific marks are sufficient, this method has little to recommend it.

Banding is most commonly, and most successfully, used with birds. Rings are applied to the legs, providing either individual-specific information detectable in the hand, or group-specific information detectable by field-observation. Mammals, too, especially larger mammals, have often been marked with tags: usually ear-tags, collars or leg-rings.

The application of a mark by tagging, banding or mutilation is a delicate procedure, and should only be attempted by people with the appropriate training. Needless to say, cruelty to the experimental animals should be minimized whatever the species and method.

5.3.5 Radioactive labels

Radioactive isotopes, applied either internally or externally, are now commonly used to mark animals. The relative lack of variety, however, as well as the considerations of danger and expense, mean that this method is of comparatively little use for capture-recapture studies.

5.4 Release

The ideal situation concerning the release of individuals is when an animal is replaced, unaffected, at its exact position prior to capture, almost immediately after being caught. Capture, handling and marking are all important in this context, of course, but the method of release is also worth considering.

For an individual to be unaffected it must not only be unharmed, but also in the same behavioural state as it was prior to capture. Release of animals still suffering from the after-effects of anaesthesia should, therefore, be avoided. So, too, should the release of animals during a period of inactivity if their capture took place during an activity period, because this may condemn them, until activity restarts, to a location they would never normally occupy. It is much better to capture, mark *and* release individuals during the same activity period. Conversely, animals may become excited during capture, making them hyperactive on release. If this is suspected, it may be desirable to give animals a short period in the environment, prior to release, in which they are constrained by some sort of cage.

Replacement of an animal at its location prior to capture is easy if

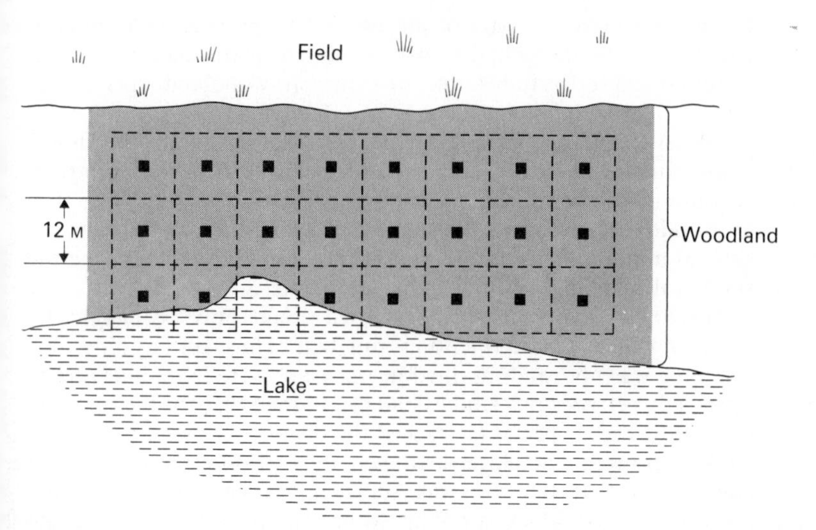

Fig. 5.2 *Drosophila* study-site with trap positions marked (■).

this is the same as the point of capture itself. This is the case whenever capture depends on the activity of the investigator. When capture depends on the activity of the animal itself, however, there are two alternatives. The animals from a trap can either be released at the location of the trap itself, on the assumption that, their true prior position being unknown, the trap's location represents their 'mean' position; or they can be released at random throughout that trap's area of influence. The disadvantages of the second alternative are the time it takes, and the fact that each individual will probably be replaced at a position quite different from that prior to capture. The advantages are that individuals *as a group* will be redistributed approximately as before, and unnatural over-crowding will be avoided.

5.5 Case studies

The following three examples have been chosen, not because they represent perfection or anything near it, but because they exemplify the problems facing real investigators in real situations. Thus, wherever possible, the constraints of the particular circumstances have been pointed out, and the twin aims of maximizing biological and statistical reliability stressed.

5.5.1 Drosophila

The fruit-fly, *Drosophila*, is an animal which has long been favoured by the geneticist. As part of an attempt to provide an ecological background to the genetic work, I, myself, performed a series of capture-recapture studies on two British woodland species: *D. obscura* and *D. subobscura*.

The site, and the arrangements of traps within it, are shown diagrammatically in Fig. 5.2. The site itself was located in a narrow corridor of woodland, because both species show a marked preference for wooded environments. This meant that migration to and from the site was minimized, recapture rates were comparatively high, and the statistical reliability of the results acceptable.

The adult *Drosophila* being studied could only be caught by attracting them to a fermenting-fruit bait. Thus, on any one day, sampling was not random: flied nearer traps were more likely to be attracted. Yet, previous studies had established that the flies' powers of undisturbed dispersal far exceed the distance between traps. The bait-boxes were, therefore, arranged in a grid-formation, on the assumption that the site was sufficiently homogeneous for the position of flies, *before* sampling, to be random with respect to the location of the traps. This was aided by removing the traps between samples. There ought, therefore, to have been little *inherent* variation in catchability; and, for the purposes of capture-recapture, the sampling could be considered random.

The actual size and spacing of the boxes was also important. The greater the surface area of the bait, the more attractive it will be. Large bait-boxes will, therefore, ensure sufficient data; but they will also attract flies from beyond the experimental site, leading to unnaturally high densities. Consequently, small traps (100 mm × 100 mm × 75 mm perspex boxes, one-quarter filled) were preferred. Their number was constrained by severe limitations in the number of available personnel: one 'biologist' and one occasional helper. Moreover, their spacing had to be such that there were neither unattractive 'holes' in the site (large site area – low efficiency of capture – low statistical reliability); nor areas where adjacent traps were in severe competition for flies (high statistical reliability – small, perhaps atypical, site). The distances chosen obviously had to be a compromise between the statistical and biological demands. In fact, the answer had to be determined by a quite separate study of the relationship between spacing and trap-efficiency.

Having designed the trap-arrangement satisfactorily, the next decision concerned the timing of the trapping process. The activity rhythms of the two species effectively limit trapping to the hours preceding dusk, and the one hour immediately before is often the most efficient. This maximal efficiency was sacrificed, however, because flies must be caught, marked and released, and still have time to remix with the population before nightfall. Moreover, it was important, as always, that sampling time should be short in comparison with total time. The trapping period was, therefore, limited to a ninety-minute period at least ninety minutes before activity ceased.

Trapping itself involved simply approaching the bait-box stealthily, and putting on the lid. Carbon dioxide was then injected into the box, anaesthetizing the flies. These were then aspirated from the bait surface. This procedure was usually successful in that a large proportion of the flies attracted were caught, and very few of those caught were harmed. *Drosophila* recover very quickly from carbon dioxide, apparently without any ill-effect.

The captured flies were then examined and recorded. Male and female *Drosophila* are distinguishable in the field with a watchmaker's eye-piece, and were treated quite separately throughout the study. (The subsequent discovery of many sex-related differences testified to the wisdom of this.) Unfortunately, the two species are not so easily distinguished, and the following procedure was, therefore, adopted. On each day a random, 10% sample of the catch was retained. These flies, as well as the whole final-day catch, were then returned to the laboratory and accurately identified. The overall proportions obtained were then used to infer the number of each species caught previously. For instance, if there were, overall, equal numbers of both species of male, then a day *2* catch of 100 males would be recorded as '50 *D. obscura* and 50 *D. subobscura*'. This was the only way of overcoming the difficulties associated with the study of two closely-related and virtually indistinguishable species.

Of course, when the flies were examined in the field the number of recaptures was also noted. However, because the numbers were comparatively low, the above procedure for species-partitioning was considered to be too inexact. Instead, all recaptures were returned to the laboratory and identified unequivocably. Thus, the inherent difficulties presented by these species necessitated special procedures in the field, and a minor alteration in the underlying model.

Flies were marked with a micronized dust which fluoresced in ultra-violet light; marks were searched for with a portable lamp, under a black-out sheet. Newly-aspirated flies where transferred – anaesthetized – to a dry tube which was coated with a fine layer of dust. They were rolled in this for approximately ten seconds, and then transferred to another dry 'cleaning' tube. There they quickly recovered, and removed the dust from most of their body – with the important exception of the dust around their thorax which they were unable to reach. This procedure undoubtedly marked all flies in a way which would make them recognizable on recapture. But, despite the fact that many marked individuals flew off less than five minutes after capture, the several stages of handling threw doubt on the assumption that the flies were unaffected. Consequently, any flies which failed to fly off immediately were left *in situ* at the bottom of the upright tube. If they were still present on the following day, they were assumed to have been harmed and could be recorded as 'not released'. Thus, the procedure aimed to mark flies efficiently without harming them – and provided a means of discerning when this aim was not realised.

The limitations in personnel, combined with the desirability of brief samples, meant that it would have been impracticable to deal with each of the twenty-three traps individually. Instead they were divided into eight 'rows'. Flies were then aspirated and recorded according to row, and marking occurred in only eight tubes. Releases took place (and tubes were left) at the mid-point of each row. Consequently, no fly was released more than 12 m from its point of capture.

Finally, the limitations in the numbers of dust colours available, meant that each study was restricted to five consecutive days. This, the efficiency of capture, and the apparent validity of the assumption of a constant survival-rate, all combined to suggest the Fisher-Ford method as the most appropriate.

5.5.2 Juvenile salmonids

The requirements of both fisheries managers and sportsmen have led to considerable interest in estimating the population size of salmonids. The following procedure for doing so by capture-recapture has been developed at the Freshwater Fisheries Unit at Liverpool, and is used as a class exercise under the supervision of Dr. K. O'Hara. The populations concerned are of juvenile salmon and trout in the River Mynach: a small nursery feeder stream of the River Dee.

Having chosen a section for study, it is possible, because the river is only about two metres wide, to arrange 'stop nets' at either end, effectively closing the population. Recapture-rates and statistical precision are then increased, and the simpler models can be used. It is known, however, that juvenile salmonids have a rather narrow home range. It is probably better, therefore, to leave the population in its naturally open state. This leads to a relatively minor loss of precision, but an important gain in realism.

Fish are captured using electric fishing equipment, with which there is a choice between alternating and direct current. Alternating current has a longer range and actually stuns the fish; but it may also be harmful both to the investigators (if they accidentally enter the water), and to the fish themselves. Direct current, on the other hand, is safe to use, and merely attracts fish to the hand-held anode. Direct current is, therefore, preferred. As the anode is moved along the section, attracted fish are captured with a hand net. They are then placed temporarily in a bucket, and transferred to a keep-pen, in batches, before the bucket water becomes deoxygenated. (In order to avoid affecting the fish a second time, care is taken to ensure that this keep-pen is never in the part of the river being fished.) There is a known bias towards larger fish with electric fishing, and probably a similar bias inherent in investigators when attracted fish are being netted. Consequently, all fish are measured (see below), and data from different age-classes analysed separately.

When the whole section has been fished, it is necessary to handle, examine and mark each individual. For obvious reasons the major considerations are speed and the avoidance of damage. Fish are transferred in batches from the pen to another bucket. The number in each batch is dependent of temperature: the higher this is, the more active the fish are; this in turn leads to higher oxygen consumption, and, therefore, to smaller batches in each fresh bucket. Salmonids are difficult to handle and easy to damage – especially in warm weather. For this reason, fish are next narcotized, a few at a time, in a bucket of the fish-anaesthetic, MS–222. They are then examined individually, and the adipose fin of unmarked individuals is clipped. This is a non-specific mark, easily recognized subsequently, which, on recapture, informs the investigator that the fish also carries a more informative mark. This is in the form of a date-specific brand, given by a freeze-brander cooled with liquid nitrogen. The brand itself causes very little damage, especially as the lateral line is avoided; but it is during the subsequent few days that there is a pigment-cell reaction to this brand, which ultimately develops into a 'mark'. It is this development time which necessitates the clipping of the adipose fin. Developing marks are noticeable to an investigator who is alerted by the presence of a clipped fin, but may be missed by one who is unprepared.

Having examined an individual for previous marks, and marked it again, it is necessary to identify the fish, and to measure it. In populations with a wide range of age-classes, measurement is insufficient as a means of ageing older individuals; and if age-classes are to be analysed separately – as they should be – scales must be removed and examined. In the River Mynach, however, all of the salmon and most of the trout are in their first or second year. Size-classes are, therefore, indicative of age – although there are several size-classes within each age-class. Measurement is made easier, and therefore quicker, by the use of a measuring board. This is simply a sheet of paper on a wooden board with separate longitudinal sections for each species and mark-group. Fish are placed on the appropriate section with their head against a line at one end. Their length is then marked at the fork of the tail with a pin-prick on the paper. Actual measurement can then be carried out after all the fish have been released.

In this way, individuals are rapidly returned to a fresh bucket, where their recovery from handling and the effects of MS–222 can be monitored. This is usually complete within a few minutes. Individuals which have recovered are then placed in a fresh keep-pen. When all individuals have been examined, marked and placed here, they are examined once again for any ill-effects; and fish which at any stage appear to be harmed are retained. Finally, the individuals in the keep-pen are carefully and gently released, a few at a time, in the centre of the section. Even at this stage, any individuals with apparently aberrant behaviour are recaught and retained.

It can be seen, overall, that the removal of fish from their natural millieu in a capture-recapture study, necessitates considerable caution and speed during handling, examining, marking and releasing. It can also be seen that the specific ways in which caution and speed are ensured, are determined by a detailed understanding of the particular populations concerned.

5.5.3 Brown rats

The enthusiasm which greeted the introduction of the anti-coagulant poison warfarin, as a means of controlling the brown rat, *Rattus norvegicus*, was tempered around 1960 by the discovery of resistant populations. It became obvious from that time that the control of brown rats, like the control of other pests, was a complicated venture demanding detailed knowledge of the species concerned. Thus it was that Bishop and Hartley (1976) set out to discover '..... something of the size, structure and degree of movement within and between populations'. Part of their work was a capture-recapture study.

The area they chose was in agricultural land near Welshpool, mid-Wales. There, their initial understandable desire to deal with a naturally 'closed' population suggested an area of about 30 km^2 surrounded by moorland. This, however, was far too large to be practicable. Instead, they had to concentrate on an area of about 1 km^2, enclosed to some extent by two rivers. Within this area, there were rats both in farm buildings, and in the nearby hedgerows and woodlands. Farmers were opposed, however, to capturing rats only to release them again. This meant that the capture-recapture studies had to virtually ignore the rats inside the buildings.

Preliminary trapping showed that the populations of rats were concentrated in particular parts of the hedgerows and woodlands. (Elsewhere, the traps were sprung by robins, baby rabbits, woodmice, coots and grey squirrels.) Consequently, 122 permanent trapping stations were established at irregularly-spaced sites. Such spacing allowed the traps to be concentrated in those areas where infestation appeared to be heaviest. There were three trap lines, and, within each, spacing varied between 1 m and 30 m. The stations were permanent so as to minimize the effect of 'new object reaction'.

Having established their stations, Bishop and Hartley had to determine their sampling procedures. They used steel-mesh cages 42 cm × 15 cm × 11.5 cm, baited with wheat, maize and meal, with a spring-loaded door released by a treadle near the aluminium bait tray. These traps combined lightness and cheapness with strength (tractors and cows were the main hazards), and allowed the examination of sprung traps before emptying. In order to protect captive animals from the weather, the traps were either wrapped in polythene sheets or placed under galvanized iron gutters.

Sprung traps were emptied into a cloth bag, and rats weighed with a spring balance. They were then transferred to a 'constraining cylinder' made of wire mesh, where they were examined, sexed, and, if captured for the first time, individually marked by clipping off two toes. This procedure was important, in that individually-marked rats could be repeatedly examined, and changes in their physical condition noted. Also, if the mark status of an individual was in some doubt, (perhaps because of a 'natural' mutilation of one of its toes), then its sex and physical condition could be determined and any uncertainty removed.

Trapping took place over a period of two years. At the beginning of each month the traps were pre-baited for five nights (as a counter to trap-shyness), set for three nights, baited unset for three nights, and then set for a final night. This procedure was adopted in order to catch rats actually infesting a hedge, without attracting too many from other areas or sustaining them wholly on bait. Bishop and Hartley also hoped that the sequence of trap-nights would lead to Fisher-Ford estimates (for males and females separately) over a period of a week; or, failing that, a Petersen estimate between days *3* and *7*.

Examination of the trapping data, however, suggested that within each month the rat populations were exhibiting both trap avoidance *and* trap addiction. Despite the precautions taken, therefore, the data were unsuitable for capture-recapture analysis; and the study, judged by its own *a priori* criteria, had failed – as so many studies do.

Bishop and Hartley could at this stage have ignored the behavioural aberrations and continued with their analysis; or they could have abandoned the whole project as unsatisfactory. But instead they adopted a third alternative. They *used* their data to modify their method of analysis. Examination of the recapture histories suggested that, although trap avoidance and addiction developed within each month, these had disappeared by the next month. It seemed sensible, therefore, to combine each group of four samples into *one* monthly sample. In other words, rats which had been captured once, twice or even more times within a month, were only considered to have been captured once. Thus, trap avoidance and trap addiction *within* a month became irrelevant. Moreover, the sampling intensities for each (monthly) sample were elevated. Accuracy and reliability had, therefore, been gained at the expense of the ability to monitor short-term changes in abundance. Nevertheless, as a result of these modifications, it was possible to apply Jolly's method to one of the 'populations' at least.

In retrospect, Bishop and Hartley's experience shows that initial hopes (of examining a closed population, or of studying *all* individuals within a study-site) are frequently not fulfilled. It also shows that biological insight and common sense are, in themselves, not enough to ensure satisfactory capture-recapture data. But, more than anything else, it shows that, even when initial attempts are thwarted, the resourceful and thoughtful investigator can often get useful and meaningful information from his data.

References

Further reading is likely to be required in two general areas. (a) References giving fuller descriptions of practical techniques are marked with an asterisk below. (b) Other models and rigorous statistical derivations are to be found in Seber (1973).

Andersen, J. (1962). Roe-deer census and population analysis by means of modified marking and release technique. In: *The Exploitation of Natural Animal Populations*. Eds. Le Cren, E. D. and Holdgate, M. W. Blackwell, Oxford.

Bailey, N. T. J. (1951). On estimating the size of mobile populations from capture-recapture data. *Biometrika*, **38**, 293–306.

Bishop, J. A. and Hartley, D. J. (1976). The size and age structure of rural populations of *Rattus norvegicus* containing individuals resistant to the anti-coagulant poison warfarin. *J. Anim. Ecol.*, **45**, 623–646.

Bishop, J. A. and Sheppard, P. M. (1973). An evaluation of two capture-recapture models using the technique of computer simulation. In: *The Mathematical Theory of the Dynamics of Biological Populations*. Eds. Bartlett, M. S. and Hiorns, R. W. Academic Press, London and New York.

Caughley, G. (1977). *Analysis of Vertebrate Populations*. Wiley, London and New York.

Cormack, R. M. (1973). Commonsense estimates from capture-recapture studies. In: *The Mathematical Theory of the Dynamics of Biological Populations*. Eds. Bartlett, M. S. and Hiorns, R. W. Academic Press, London and New York.

*Delany, M. J. (1974). *The Ecology of Small Mammals*. Studies in Biology no. 51. Edward Arnold, London.

*Dobson, R. M. (1962). Marking techniques and their application to small terrestrial animals. In: *Progress in Soil Zoology*. Ed. Murphy, P. W. Butterworths, London.

Dowdeswell, W. H., Fisher, R. A. and Ford, E. B. (1940). The quantitative study of populations in the Lepidoptera. 1. *Polyommatus icarus* (Rott). *Ann. Eugen.*, **10**, 123–136.

Fisher, R. A. and Ford, E. B. (1947). The spread of a gene in natural conditions in a colony of the moth *Panaxia dominula* (L). *Heredity*, **1**, 143–174.

Gilbert, N. (1973). *Biometrical Interpretation*. University Press, Oxford.

Jackson, C. H. N. (1937). Some new methods in the study of *Glossina morsitans*. *Proc. Zool. Soc. London*, 1936, 811–896.

Jackson. C. H. N. (1939). The analysis of an animal population. *J. Anim. Ecol.*, **8**, 238–246.

Jolly, G. M. (1965). Explicit estimates from capture-recapture data with both death and immigration – stochastic model. *Biometrika*, **52**, 225–247.

Kettlewell, H. B. D., Berry, R. J., Cadbury, C. J. and Phillips, G. C. (1969). Differences in behaviour, dominance and survival within a cline. *Heredity*, **24**, 15–25.

*Lagler, K. F. (1971). Capture, sampling and examination of fishes. In: *Methods for Assessment of Fish Production in Fresh Waters.* Ed. Ricker, W. E. Blackwell, Oxford.

Leslie, P. H. (1958). Statistical appendix. *J. Anim. Ecol.*, **27**, 84–86.

Lincoln, F. C. (1930). Calculating waterfowl abundance on the basis of banding returns. *U.S. Dept. Agric. Circ.*, **118**, 1–4.

Manly, B. F. J. (1970). A simulation study of animal population estimation using the capture-recapture method. *J. appl. Ecol.*, **7**, 13–39.

Manly, B. F. J. (1971). A simulation study of Jolly's method for analysing capture-recapture data. *Biometrics*, **27**, 415–424.

Manly, B. F. J. (1973). A note on the estimation of selective values from recaptures of marked animals when selection pressures remain constant over time. *Res. Popul. Ecol.*, **14**, 151–158.

Manly, B. F. J. and Parr, M. J. (1968). A new method of estimating population size, survivorship, and birth-rate from capture-recapture data. *Trans. Soc. Brit. Ent.*, **18**, 81–89.

Morejohn, G. V. and Howard, W. E. (1956). Molt in the pocket gopher. *Thomomys bottae. J. Mammal.*, **37**, 201–213.

Parker, R. E. (1979). *Introductory Statistics for Biology.* Second edition. Studies in Biology no. 43. Edward Arnold, London.

Petersen, C. G. J. (1896). The yearly immigration of young plaice into Limfjord from the German sea, etc. *Rept. Danish Biol. Stn.*, **6**, 1–48.

Robson, D. S. and Regier, H. A. (1964). Sample size in Petersen mark recapture experiments. *Trans. Amer. Fish. Soc.*, **93**, 215–226.

Roff, D. A. (1973a). An examination of some statistical tests used in the analysis of mark-recapture data. *Oecologia* (*Berl.*), **12**, 35–54.

Roff, D. A. (1973b). On the accuracy of some mark-recapture estimators. *Oecologia* (*Berl.*), **12**, 15–34.

Seber, G. A. F. (1973). *The Estimation of Animal Abundance and Related Parameters.* Griffin, London.

Sheppard, P. M., Macdonald, W. W., Tonn, R. J. and Grab, B. (1969). The dynamics of an adult population of *Aedes aegypti* in relation to dengue haemorrhagic fever in Bangkok. *J. Anim. Ecol.*, **28**, 661–702.

*Southwood, T. R. E. (1966). *Ecological Methods.* Methuen, London.

*Stott, B. (1971). Marking and tagging. In: *Methods of Assessment of Fish Production in Fresh Waters.* Ed. Ricker W. E. Blackwell, Oxford.

*Taber, R. D. and Cowan, I. M. (1969). Capturing and marking wild animals. In: *Wildlife Management Techniques.* Ed. Giles, R. H., Jr. The Wildlife Society, Washington.

*Twigg, G. I. (1975a). Catching mammals. *Mammal Rev.*, **5**, 83–100.

*Twigg, G. I. (1975b). Marking mammals. *Mammal Rev.*, **5**, 101–116.

White, E. G. (1975). Identifying population units that comply with capture-recapture assumptions in an open community of alpine grasshoppers. *Res. Popul. Ecol.*, **16**, 153–187.

*Woodbury, A. M. (1956). Uses of marking animals in ecological studies: marking amphibians and reptiles. *Ecology*, **37**, 670–674.

Wright, S. (1969). *Evolution and the Genetics of Populations. Volume II. The Theory of Gene Frequencies.* The University of Chicago Press, Chicago.